Basic Electrophysiological Methods

Basic Electrophysiological Methods

EDITED BY

Ellen Covey, PhD

Department of Psychology
University of Washington
Seattle, WA

Matt Carter, PhD

Department of Biology
Williams College
Williamstown, MA

OXFORD
UNIVERSITY PRESS

OXFORD
UNIVERSITY PRESS

Oxford University Press is a department of the University of
Oxford. It furthers the University's objective of excellence in research,
scholarship, and education by publishing worldwide.

Oxford   New York
Auckland   Cape Town   Dar es Salaam   Hong Kong   Karachi
Kuala Lumpur   Madrid   Melbourne   Mexico City   Nairobi
New Delhi   Shanghai   Taipei   Toronto

With offices in
Argentina   Austria   Brazil   Chile   Czech Republic   France   Greece
Guatemala   Hungary   Italy   Japan   Poland   Portugal   Singapore
South Korea   Switzerland   Thailand   Turkey   Ukraine   Vietnam

Oxford is a registered trademark of Oxford University Press
in the UK and certain other countries.

Published in the United States of America by
Oxford University Press
198 Madison Avenue, New York, NY 10016

Library of Congress Cataloging-in-Publication Data
CIP to come
ISBN 978–0–19–993980–0

The science of medicine is a rapidly changing field. As new research and clinical experience broaden our
knowledge, changes in treatment and drug therapy occur. The author and publisher of this work have checked
with sources believed to be reliable in their efforts to provide information that is accurate and complete, and
in accordance with the standards accepted at the time of publication. However, in light of the possibility of
human error or changes in the practice of medicine, neither the author, nor the publisher, nor any other party
who has been involved in the preparation or publication of this work warrants that the information contained
herein is in every respect accurate or complete. Readers are encouraged to confirm the information contained
herein with other reliable sources, and are strongly advised to check the product information sheet provided
by the pharmaceutical company for each drug they plan to administer.

9 8 7 6 5 4 3 2 1
Printed in the United States of America
on acid-free paper

Contents

Contributors

Antoine Adamantidis
Department of Neurology, Inselspital,
Bern University Hospital
Freiburgstrasse, Bern, Switzerland
Department of Psychiatry,
McGill University
Douglas Mental Health University
Institute
Montreal, Canada

Monica M. Arnold
Department of Psychiatry and
Behavioral Sciences and Department of
Pharmacology, University of Washington
Seattle, WA

Lauren M. Burgeno
Department of Psychiatry and
Behavioral Sciences and Department of
Pharmacology, University of Washington
Seattle, WA

R. Michael Burger
Department of Biological Sciences,
Lehigh University
Bethlehem, PA

Luke Campagnola
Department of Otolaryngology/Head
and Neck Surgery and Curriculum
in Neurobiology, University of North
Carolina at Chapel Hill
Chapel Hill, NC

Matt Carter
Department of Biology, Williams College
Williamstown, MA

William L. Coleman
Department of Biological and Allied
Health Sciences
Hartline Science Center, Bloomsburg
University of Pennsylvania
Bloomsburg, PA

Ellen Covey
Department of Psychology,
University of Washington
Seattle, WA

David Ferster
Department of Neurobiology,
Northwestern University
Evanston, IL

William Frost
Department of Cell Biology and
Anatomy, The Chicago Medical School
Rosalind Franklin University of
Medicine and Science
Chicago, IL

Xue Han
Department of Biomedical Engineering,
Boston University
Boston, MA

Todd C. Handy
Department of Psychology, University of
British Columbia
Vancouver, Canada

Sonia Jego
Integrated Program in Neurosciences,
McGill University
Montreal, Canada

Julia W. Y. Kam
Department of Psychology, University of
British Columbia
Vancouver, Canada

Caleb Kemere
Department of Electrical and Computer
Engineering, Rice University
Houston, TX

Richie E. Kohman
Department of Biomedical Engineering,
Boston University
Boston, MA

Paul B. Manis
Department of Otolaryngology/Head
and Neck Surgery and Curriculum
in Neurobiology, University of North
Carolina at Chapel Hill
Chapel Hill, NC

Paul E. M. Phillips
Department of Psychiatry and
Behavioral Sciences and Department
of Pharmacology, University of
Washington
Seattle, WA

Sean Reed
Integrated Program in Neurosciences,
McGill University
Montreal, Canada

Samantha R. Summerson
Department of Electrical and Computer
Engineering, Rice University
Houston, TX

Hua-an Tseng
Department of Biomedical Engineering,
Boston University
Boston, MA

Jian-young Wu
Department of Neuroscience,
Georgetown University Medical Center
Washington D.C.

Introduction

Matt Carter and Ellen Covey

Electrophysiology is the branch of neuroscience that investigates the electrical activity of cells and tissues, as well as the molecular and cellular processes that govern their signaling. Because the function of the nervous system is ultimately rooted in the electrical activity of neurons, electrophysiology techniques are widely considered to be the backbone of neuroscience research, even in the modern era of genomics, optics, and imaging. Indeed, recently developed, cutting-edge neuroscience techniques can only reach their full potential when combined with classical electrophysiological methods.

"Electrophysiology" is really an umbrella term for an assortment of diverse techniques, each of which addresses specific questions and requires specific instruments, tissue preparations, methods of analysis, and expertise. For example, the electrophysiological methods employed by a systems neuroscientist attempting to record the receptive fields of neurons in awake, behaving animals are very different than the methods used by a molecular neuroscientist attempting to determine the contributions of a single ion channel to synaptic potentiation. The ultimate goal of this book is to differentiate between the major categories of electrophysiology techniques, describing specific applications, protocols, instruments, and frequently asked questions for each.

The book is not meant to be a stand-alone guide that will enable you to follow a set of step-by-step instructions and implement a particular technique immediately. Instead, it provides an overview of the theory behind each technique so that the reader understands the basic principles of how it works, a discussion of what it can and cannot do, and an illustrative description of the equipment, procedures, and analysis for a typical application. Each technique can be used in a variety of model systems, for a variety of purposes, so it is ultimately up to the user to adapt it in whatever ways are appropriate. It is highly recommended that anyone who intends to use a technique arrange to work with someone who can provide a hands-on demonstration of the technique in action, and experience with every aspect of using it. This book is intended to prepare the reader for hands-on learning, reducing the number of questions, and facilitating rapid acquisition of the necessary knowledge and skills.

Perhaps the greatest difference among the various electrophysiology techniques lies in the location of the probe used to measure electrical signals relative to the neuron or neurons of interest. For example, the optimal method of recording from a single neuron is to place a glass micropipette directly adjacent to a small area (or "patch") of the plasma membrane, forming a tight connection between electrode and cell. If gentle suction is applied to the distal end of the micropipette, the connection between pipette and cell becomes so tight that a very high resistance is achieved and microscale, sub-threshold electrical signals can be measured. Chapter 1: Patch Clamp Recording in Brain Slices describes the principles and practice of various "patch clamp" techniques, which can ultimately provide information about the current that flows through single ion channels, the effects of ions and pharmacological agents on cell physiology, and the relative physiological properties of different categories of neurons. Patch clamp techniques are typically performed *in vitro* such that the investigator can lower the glass micropipette directly onto a neuron of interest under visual guidance using a microscope. Although technically challenging, it is also possible to use patch clamp techniques in an anesthetized animal. Chapter 2: Patch Clamp Recording *in vivo* describes the technology and procedures required to patch onto a neuron within the intact brain, obtaining the same type of information that is available *in vitro*.

Electrical probes can also be placed a short distance outside of neurons in the extracellular environment. The changes in membrane voltage that occur during an action potential generate local differences in potential on the outer surface of an active neuron. Therefore, action potentials can be detected in the extracellular space near the membrane of an active neuron by measuring the potential difference between the tip of a recording electrode relative to a ground electrode. Chapter 3: Extracellular Single-Unit Recordings and Neuropharmacological Methods describes the methodology of extracellular electrophysiology, which is the approach most often used in *in vivo* preparations. These techniques can address questions such as to how an individual neuron or small group of neurons encodes information, the role of a neuron in a specific sensory, motor, or cognitive operation, or the correlation of neural activity in one brain region with activity in another brain region. It is also common to combine extracellular recording with other methods such as local application of pharmacological agents to investigate the role of specific classes of receptors, ion channels, etc. These methods are described in Chapter 3.

Chapter 4: Multi-Electrode Recording of Neural Activity in Awake Behaving Animals describes the use of multiple extracellular electrodes in the same preparation to assess the neural activity of many neurons at the same time. This method is practical for use in awake, behaving animals. Multi-electrode techniques are useful to show how an entire population of neurons responds to a stimulus, potentially revealing temporal relationships among different neurons within the group and the relation of activity to stimuli and/or behaviors of interest.

Specialized extracellular electrodes allow for the measurement of specific neurochemical events in the brain. These neurochemicals typically include monoamine neurotransmitters such as serotonin, norepinephrine, and especially dopamine. A carbon fiber microelectrode is inserted into the brain and a specific voltage is applied. When the monoamines encounter the surface of the electrode, they undergo an oxidation reaction, producing a measurable change in current. The magnitude of this current is proportional to the number of molecules oxidized, therefore allowing for the precise measurement of neurochemicals on a physiological timescale. Chapter 5: Amperometry and Voltammetry Fast-Scan Cyclic Voltammetry in Behaving Animals describes these neurochemical measurement techniques including how to read and analyze the waveforms produced by oxidation and reduction states resulting from fluctuations in neurochemical activity.

Other electrophysiological recording techniques place an extracellular electrode in a location such that the aggregate activity of many hundreds or thousands of neural elements is recorded simultaneously. Local field potential (LFP) recordings are obtained from low-impedance electrodes placed within the brain, and are dominated by the electrical current flowing from the summation of all dendritic synaptic activity within a local volume of tissue. In an electroencephalogram (EEG), the electrode is placed on the surface of the scalp to record the voltage fluctuations from many thousands of neural elements at once, providing information about the synchronization of neural activity, mainly in the cortex. Chapter 6: Electroencephalography and Related Techniques Recording in Humans focuses on the application of EEG to human subjects and describes the acquisition and analysis of signals from multiple sources on the scalp. Because EEG is relatively noninvasive, it is the main electrophysiological method applied to humans. However, it can also be a useful tool in animal studies. Chapter 7: Electroencephalography and Local Field Potentials in Animals describes the application of EEG and LFP to animal models, especially rodents. Such techniques are often used to study epilepsy, sleep, and attention in experimentally tractable animal models.

Recent advances in optics and chemical engineering have allowed the electrical activity of neurons to be visualized instead of directly measured with a recording electrode. Specialized dyes that change their fluorescence in response to changes in membrane voltage can reliably report neural activity, sometimes with a resolution of single action potentials. These techniques can be used *in vivo* and *in vitro* to reveal the electrical activity of an entire population of neurons rather than one neuron or group of neurons at a time. Chapter 8: Voltage-Sensitive Dye Imaging discusses the application and utility of these techniques for visualizing electrical activity in neurons, both individually and collectively.

Perhaps the most impactful advance in neuroscience methodology in recent years has been the development of optogenetics, a technology that allows scientists the ability to manipulate neural activity with unprecedented spatial and temporal precision. Chapter 9: Optogenetics and Electrophysiology describes how investigators can genetically target optogenetic actuators to specific neurons of interest, then manipulate activity with millisecond precision to induce or inhibit neural firing at a resolution of single action potentials. Importantly, these techniques can be combined with conventional electrophysiology techniques to simultaneously stimulate and record from neurons *in vivo* or *in vitro*.

The techniques described above and in the forthcoming chapters differ so greatly from each other that it is amazing that the major practitioners of each can all be considered "electrophysiologists." The authors of the subsequent chapters have done a remarkable job defining and characterizing exactly what each technique can be used for and which questions they can address. Furthermore, they explain the fine details of each technique for readers interested in actually purchasing equipment, preparing specimens, and collecting and analyzing data. Therefore, we hope this book provides an appreciation for the diversity of basic electrophysiology methods as well as a running start for readers interested in pursuing electrophysiology experiments in their own laboratories.

Patch Clamp Recording in Brain Slices

Luke Campagnola and Paul B. Manis

Introduction

Neurons use electrical signals to process and transmit information. These electrical signals are part of the "currency" of neural processing and appear in multiple forms that can be recorded in a variety of ways. The simplest recordings are measurements of extracellular potentials, which are generated by current flow between different parts of individual neurons. These recordings can reflect either the spiking of neurons (as in single-unit or multi-unit recordings) or field potentials that are composed of a mixture of synaptic currents and cell spiking near the recording electrode. Neither of these types of recordings requires access to the interior of the cells.

The electrical signals recorded outside of single neurons are generally very small, in the range of µV to a few mV. While these are useful for understanding the coding of information in spike trains, or the spatial and temporal organization of synaptic inputs in laminar structures, they do not readily reveal the underlying mechanisms of synaptic integration or spike generation. A modern electrophysiological method called *patch clamp* permits the measurement of transmembrane voltage and current with high resolution and low noise. What sets patch clamp apart from other methods, such as sharp-electrode intracellular recording, is its use of a tight seal between the recording electrode and the cell membrane. This seal essentially blocks external currents associated with the electrical activity of surrounding cells, allowing the experimenter to record even small subthreshold events in a single neuron. The development of these techniques to create high-resistance seals with the micrometer-sized pipette tips needed to record from small vertebrate neurons (Hamill et al. 1981, Neher et al. 1978) revolutionized electrophysiology.

There are several variants of the patch clamp method. The most commonly used method is best described as *whole-cell recording*. Of the other variants on the patch clamp method, the next most commonly used configurations are the cell-attached patch, followed by outside-out and inside-out patch recording. In this chapter, we focus on the technical aspects involved in making whole-cell patch clamp recordings and only briefly touch on cell-attached and outside-out patches.

When studying electrical signaling in neurons, there are two widely used recording methods. *Current clamp* refers to recording the voltage across the membrane of individual cells. *Voltage clamp* refers to measuring the current that is associated with conductance changes in the membrane in response to voltage changes. Current clamp and whole-cell voltage clamp require electrical access to the interior of the cell. Both current clamp and voltage clamp can be very informative about the mechanisms by which cells fire particular patterns of action potentials or by which synaptic inputs change with time and are integrated. These recording methods also permit a variety of manipulations that can yield insight into the cellular physiology and biology of specific neurons, revealing mechanisms underlying processes such as learning, memory, decision making, hormonal regulation, the construction of motor activity patterns, and perception.

In this chapter, we focus on the use of brain slices when making patch clamp recordings. Brain slices are an *in vitro* preparation, created by sectioning fresh brain tissue (Yamamoto and McIlwain 1966). Slices offer a number of advantages for analysis of cellular mechanisms and small networks, including excellent optical access to even small cellular elements such as dendritic spines, mechanical stability, and the ability to control the extracellular environment for ionic or pharmacological manipulations. Slices have an advantage over dispersed neuronal cultures in that they can retain much of the *in vivo* network structure and connection specificity, as well as the normal complement of cells. However they have disadvantages as well. Brain slices can only be used for a few hours after preparation, and they lack many of the normal activity patterns that can be observed *in vivo*. Regardless, over the past 30 years, slices have been instrumental in deepening our understanding of cellular and synaptic physiology and have significantly contributed to understanding local neuronal networks.

Overview of the Patch Clamp Technique

The patch clamp technique achieves superior recording fidelity by creating a high-resistance seal between the pipette glass and the cell membrane, then rupturing the membrane within the lumen of the pipette to allow measurement of electrical signals internal to the cell. In the methods we describe, a thin slice of brain tissue is held in a small chamber perfused with warm, oxygenated fluid, which approximates cerebrospinal fluid. This arrangement keeps the brain slice alive for several hours after dissection, during which time the experimenter will attempt to patch neurons within the slice. The recording chamber is mounted on a microscope stage to allow the experimenter to visually identify cells within the slice. Glass patch pipettes, filled with an electrode

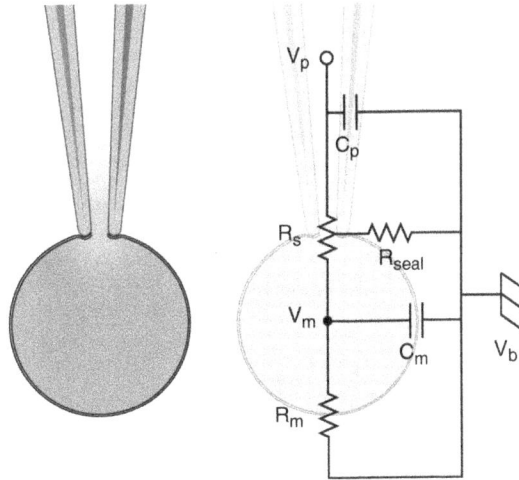

FIGURE 1.1 Schematic of glass electrode patched onto cell with equivalent circuit diagram.
V_p: Voltage inside pipette; this is the voltage controlled or measured by the amplifier, less the electro-chemical junction potential. C_p: Pipette capacitance; typically a few pF. R_s: Series (or access) resistance; this is the resistance separating the pipette from the cell body and is due mainly to the narrow pipette tip and organelles that may be blocking it. R_{seal}: Seal resistance; the resistance of the region of contact between the pipette and the membrane. To make quality recordings, this must be >1 GΩ. V_m: Membrane voltage; the voltage of the interior of the neuron relative to the bath. C_m: Cell membrane capacitance. R_m: Cell membrane resistance; also called input resistance. V_b: Bath voltage, as measured by the ground electrode.

solution approximating the composition of cytosol, are then positioned using precise micromanipulators to form a seal with the cell membrane.

By applying suction to the patch pipette, the membrane is ruptured, granting electrical access to the cell interior. However, this access is imperfect due to the electrical characteristics of the pipette. The goal of patch clamp, in the most general sense, is to measure and manipulate the voltage V_m and resistance R_m of the patched neuron (Fig. 1.1). This goal is confounded by the combination of electrical resistance at the tip of the pipette R_s and capacitance across the cell membrane C_m and pipette walls C_p. These confounds are addressed through a variety of techniques that are discussed throughout the chapter.

Equipment

A typical patch clamp recording setup is shown in Figure 1.2 and minimally consists of an amplifier, a microscope, a data acquisition system, a solution delivery system, and a chamber that holds the brain slice.

Amplifiers

Currents in neurons span a wide range, from pA for single channels and minia-ture synaptic currents to tens of nA for currents through some voltage-gated ion

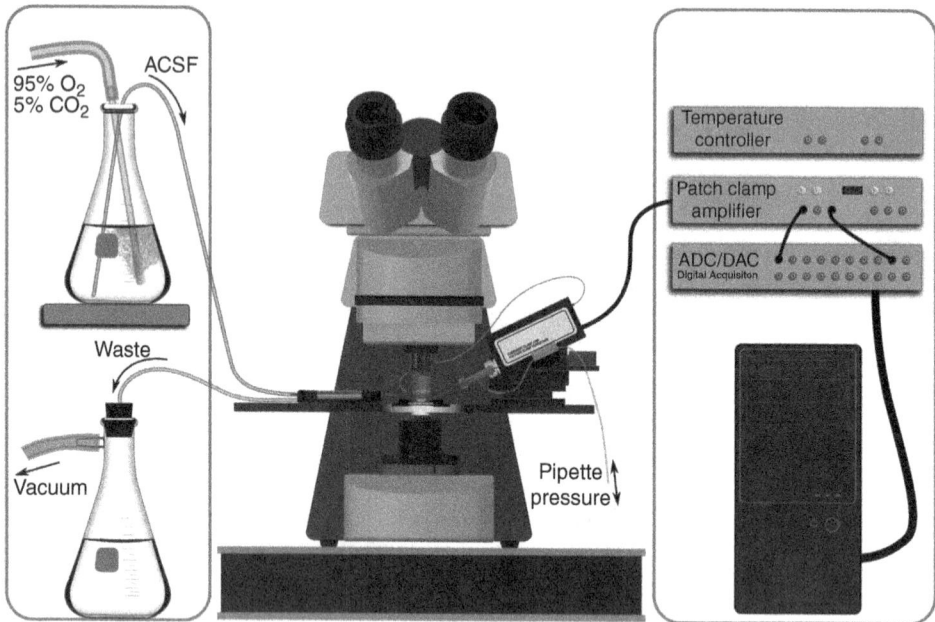

FIGURE 1.2 A minimal patch electrophysiology rig. Left to right: Oxygenated ACSF is siphoned through a fluid heater and into the recording chamber, where it continuously washes over the brain slice. Fluid is then aspirated out of the recording chamber and into a waste flask. A patch clamp amplifier headstage is mounted to a micromanipulator and holds the patch pipette, which currently impales the brain slice (detailed in Fig. 1.3). The headstage output is amplified, digitized, and finally recorded on a computer.

channels. Similarly, voltages in neurons range from tens of μV for synaptic potentials to ~100 mV for action potentials. Consequently, specialized equipment is required to amplify the signals before they are recorded. The amplifiers used in electrophysiology are uniquely suited to recording such signals while introducing as little noise as possible.

While early patch clampers had to build their amplifiers, a variety of commercial amplifiers are now available that have capabilities and specifications that would be hard for an individual laboratory to duplicate. The choice of an amplifier should be guided by the goals of the experiments, as not all amplifiers have the same merit figures for noise, or bandwidths, and some are better suited for particular experiments. The amplifiers consist of two main parts. The *headstage* is usually placed close to the recording site and may hold the electrode directly. The headstage is connected to the main part of the amplifier, which provides the controls and input and output signals, and usually interfaces with a computer. The headstage also provides a high-quality ground output that is used to ground the fluid in the recording chamber and provide a clean reference for measuring potentials.

If the goal is to make current clamp recordings, a voltage-follower headstage is recommended. This type of headstage has a very high input impedance, low input capacitance, and a fairly wide bandwidth. Usually these headstages are also equipped with a current injection circuit and circuitry that allows compensation of the electrode capacitance. As a result, they can faithfully report the voltage that appears at the input terminal, or at the wire that is inserted into the electrode. With properly prepared electrodes, the measured voltage can be nearly identical to the voltage at the tip, although it will be affected by electrode capacitance, which creates a low-pass filter that will attenuate rapid changes in voltage, as well as the voltage drop introduced by any current flowing through the electrode.

When the goal is to make voltage clamp recordings, the headstage is usually configured as a current-to-voltage converter. Because of the bandwidth limitations that arise when high gain is needed (e.g., pA to V), the amplifiers include additional internal circuitry that helps to extend the bandwidth of the amplifier. Such headstages can also be operated in a mode that allows voltage measurement (as in current clamp), but limitations of the circuit can result in distortions of the voltage waveform (Magistretti et al. 1998).

Some modern amplifiers have dual-function headstages that provide very good performance in both voltage and current clamp, making it possible to collect data in both modes from individual cells, and even to switch between modes in the middle of a sweep. For example, this can be useful if you wish to elicit an action potential or a train of action potentials, and then measure the currents associated with the ensuing after-hyperpolarization.

Many modern amplifiers offer computer control, while older amplifiers were designed in an era when computers were used only to collect data from the amplifier. Having digital control of the amplifier and some internal signal processing, such as filtering, can be advantageous since it allows the experimenter greater flexibility and can simplify the execution of particular experimental manipulations. On the other hand, the use of digital circuits, including digital signal processing, in the amplifier can introduce additional noise sources that can be difficult to eliminate and sometimes are hard to detect unless a wide-band oscilloscope is available. High-frequency noise that is above the Nyquist frequency sampling limit imposed by the rate selected for the analog-to-digital (A/D) conversion process becomes folded down into lower frequencies and adds to the apparent low-frequency noise. (The Nyquist limit is half of the A/D conversion frequency, per channel. For example, if you sample the voltage channel every 100 μsec, then the sample frequency is 10 kHz and the Nyquist limit is 5 kHz. All signals coming into the A/D converter should be low-pass–filtered below 5 kHz. In practice, for 5-kHz filtering, the sampling rate should be at least 20 kHz.) We have observed high-frequency noise on the output (following internal filtering) of a digitally controlled amplifier and suggest that for recordings under conditions where low noise is demanded by the experimental measurements, analog amplifiers without internal digital processing may be preferable, or additional external filtering following the amplifier may be required.

Electrode Holders

Properly functioning and clean electrode holders are critical to the success of patch clamp recording. The holders are typically made from a polycarbonate shell and are designed to provide mechanical stability for the electrode, low capacitance, electrical insulation, and a low-noise connection to the input of the headstage (Fig. 1.3). Holders should be cleaned regularly following the manufacturer's instructions, and they should always be cleaned whenever the electrode solution gets into the holder.

The wire in the holder is usually made from silver and is coated with a thin chloride layer. The wire should be cleaned by carefully polishing with 600 grit polishing paper and washed with ethanol to remove any residue and oils. At this point, it is recommended to handle the wire with either tweezers or gloves. The chloride layer can be created by placing a cleaned wire into diluted bleach for a day or two. The last 1–2 mm of the wire should be cleaned to bare silver so that it can make good contact with the gold pin of the electrode holder (see Fig. 1.3I). In some experiments, solutions are used that have a large electrochemical potential against silver chloride, so in this case it is important to provide a bridge, usually made of agar and 3M KCl, to connect to the silver wire. Several published protocols can be found in the literature (Shao and Feldman 2007, Snyder et al. 1999). Chlorided silver wires may also be used as the headstage ground wire in lieu of commercially available ground wires with AgCl pellets attached.

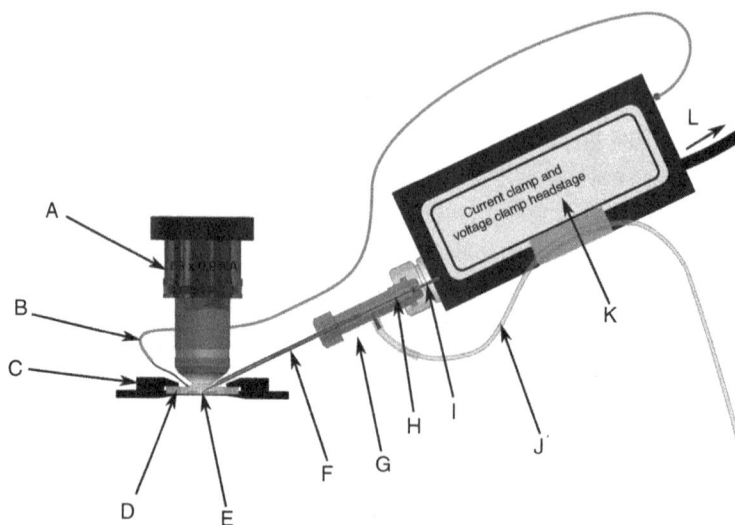

FIGURE 1.3 Patch recording equipment. A: 63× ceramic, water-immersion objective. B: Silver chloride wire connected to headstage ground output. C: Heated aluminum holder for recording chamber. D: Plastic recording chamber with glass coverslip on the bottom. E: Brain slice bathed in warm, oxygenated ACSF. F: Glass patch pipette filled with electrode solution. G: Electrode holder. H: AgCl electrode wire. This wire fits inside the patch pipette and makes electrical contact with the electrode solution as well as the (I) gold pin, which conducts electrode potential into the amplifier headstage. J: Pressure control tube. This allows the experimenter to increase or decrease the pressure inside the patch pipette. K: Amplifier headstage. L: Signal cable to amplifier.

Patching requires the application of negative and positive air pressure to the back of the pipette, so the holder will have a single port. We use a 1- to 2-cm length of Silastic tubing to provide a flexible joint, and then ~20 cm of polyethylene tubing to reach the table or the armrest on the isolation table (see Fig. 1.3J). There are several methods of delivering air pressure. Some prefer to use a mouth pipette, which provides excellent control over the pressure. We often use a 1-cc tuberculin syringe (without the needle) and an adapter to connect to the tubing. With practice, the syringe can be used to make small changes in pressure that are either slow, for making seals, or fast, for rupturing the cell membrane.

When tightening the holder onto the headstage, it is often important to be careful not to make the holder too tight, as relaxation of the Teflon coupler over time can introduce slow movement of the electrode. Similarly, the rubber retaining ring in most holders should not have much pressure on it when the cap is tightened, so that it does not relax and twist the electrode. A gentle but firm finger tightening is sufficient.

Also, there is usually a gasket that seals access to the AgCl electrode at the back of the holder. This gasket is important to provide a pneumatic seal and to keep fluid out of the connection to the headstage.

Microscope

Patch clamp recordings in brain slices are most successful when the cells to be recorded can be directly visualized, although "blind" patching is a technique that can be used under some circumstances. Direct visualization is usually achieved by using a fixed-stage, upright microscope with electrically insulating (e.g., ceramic) water-immersion objectives. The fixed microscope stage is often replaced with a fixed platform or a set of gantry towers that are fixed to the vibration isolation table, and the microscope is placed on a translatable platform. An alternate approach is to fix the microscope to the vibration isolation table and to have a translatable stage to hold the preparation and manipulators. This latter approach is commonly used when introducing laser light into the microscope (e.g., for two-photon microscopy), since the optical platform must remain well aligned with the light source. It is usually not necessary to have more than two objectives on these microscopes. A low-magnification (2.5–5×) long-working-distance objective can be used to select the region of the slice and find the electrode for coarse positioning, while a high-magnification (40× or 63×) 2- to 3-mm-working-distance water-immersion objective is needed for visualizing and patching cells. It is best if the objectives can be exchanged and returned to the same focal position without requiring refocusing the microscope, and this is achieved with a sliding objective positioner. However, it is also possible to use a rotary objective turret, although this will require refocusing for each change of objective.

For young tissue or thin preparations, the use of Nomarski differential interference contrast optics (DIC) can help with the visualization of cell membranes and fine processes such as dendrites. In addition, the use of infrared (IR) or long-wavelength illumination reduces light scattering and can be used to gain better visualization into deeper regions of the tissue. However, these long wavelengths also require a camera with good IR sensitivity (usually a CCD camera) to actually visualize the preparation. Some CCD cameras have IR filters in front of the detector that reduce their natural IR sensitivity, and these have to be removed and replaced with an appropriate IR-transmitting filter.

For older, thicker, or more heavily myelinated tissue, the use of DIC optics has little advantage, since the light polarization is partially randomized by the tissue. In this case, asymmetric or gradient illumination, followed by appropriate adjustments in image contrast on the monitor, works nearly as well. With modern cameras that have dynamic ranges of 12 or 16 bits, the contrast can be greatly increased around a mean level, allowing visualization of details that would be otherwise lost. The asymmetric or gradient illumination also helps increase the contrast. In the simplest case, such illumination can be obtained by adjusting the condenser off center from the light path or by using a high-power IR-light-emitting diode placed below the preparation and off the optical axis. We have used simple asymmetric illumination and image contrast adjustments to perform visualized patch recordings from neurons in 300-μm-thick slices of adult mouse (>80 days old) brainstem nuclei that have heavy myelination. In some cases, having software or hardware adjustments that allow the displayed image contrast to be enhanced, while subtracting background light levels, can also help visualize cells.

Blind patching is a technique whereby the patching is done without direct visualization (Blanton et al. 1989). In this case, the only feedback available is the electrical signal from the electrode. Blind patching can be done *in vivo* or in thick tissues where no visualization is possible. However, the success rate is lower than for visualized patching.

Several approaches that fundamentally consist of optical workbenches with objectives and reconfigurable mechanical arrangements are now available. These may be preferable in some situations as they allow the rig to be changed to meet the demands of specific experiments or new optical configurations much more easily than if a dedicated microscope with enclosed optics is used.

Manipulators

Positioning the electrode requires the use of manipulators that allow smooth motion in three axes at the submicrometer level. Typically, this is achieved with mechanical, hydraulic, piezoelectric, or stepping-motor manipulators. Each type has some advantages and disadvantages, but the current trend is toward piezoelectric and stepping-motor manipulators that have remote control units so that moving the electrode does not require touching the manipulator itself. The manipulators should be mounted securely on the same platform as the recording chamber. These manipulators will also often have a mechanical arrangement that allows the headstage and electrode holder to be easily brought out from under the objective to change electrodes.

Vibration Isolation

An important component of any patch clamp setup is reduction of building vibration. Most buildings have vibration that arises from air handling systems and nearby road-ways (or railroad tracks), as well as foot traffic in the hall. Vibration that is transmitted to the electrode can make patching difficult or impossible. For patch clamp recording, tables can vary in size, although we typically use 30 × 48-inch tables with 4-inch-deep tops to allow sufficient room for the microscope, light sources, and ancillary equipment that is on the table. Smaller tables can be used as well, if they are located in an area with less vibration. Larger tables are only needed if there will be additional optics, such as lasers, on the table. The tables are "floated" using nitrogen supplied through a regulator. House air systems can be used if they have sufficient pressure, but it is recommended to provide an air filter and a water trap in the system to avoid mishaps that could damage the table.

Other Hardware

Stimulators. One of the most common ways to activate pathways in a brain slice is to electrically stimulate the tissue using a bipolar or concentric electrode, usually no more than 250 µm in diameter. Simple stimulating electrodes can be made by twisting small-gauge (22–30 ga) Teflon-coated platinum wires together, cutting the ends flush with a sharp razor, attaching it to a twisted pair of wires that go to the stimulator, and inserting the platinum end through a fire-polished Pasteur pipette until the ends stick out of the pipette. A small drop of glue at the end of the pipette will help hold the wires in place. Commercial electrodes are also available in a variety of sizes and configurations from several vendors. The basic requirement for the stimulator is that the current (or voltage) and pulse duration be controlled. Typical pulse durations are 0.05–0.2 msec per stimulus. Voltages range from <1 to ~100 V or, if using constant current pulses, from tens of µA up to ~1 mA. Stimulus parameters are highly dependent upon the tissue type as well as the electrode configuration. The stimulator hardware consists of the *pulse generator* (this can be a computer or a stand-alone unit) along with an *isolation unit* that drives the electrode through a circuit that is electrically isolated from the rest of the setup. This isolation occurs either through an optical coupler or a transformer. The output of the isolation unit should not be grounded.

Pipette puller. The preparation of the patch pipettes requires a puller suited to the purpose. Modern pullers are microprocessor-based devices that can create a pair of patch pipettes from glass *blanks* by heating the glass with either a filament or a laser and cooling the glass with a jet of air. The choice of puller is not critical as long as it is easy to modify the pulling pattern of heating and cooling and force. With some pullers it may also be necessary to have a *microforge* to fire-polish the tips of the electrodes. We have not found this necessary with a laser puller.

Slicers. The preparation of brain slices requires a slicer. To minimize damage to the tissue, slicers that use a vibrating blade that can be advanced through the tissue with a controlled rate, oscillation speed and distance, and angle seem to work best. The slicer should be dedicated to brain slice preparation, as contamination with fixative or chemicals that might be encountered during histological processing is not conducive to the preparation of healthy, living brain slices.

Water filtration system. It is extremely important to have high-quality water when preparing solutions for brain slices and patch clamp recordings. Contamination of the water used to make solutions by water treatment chemicals, bacteria, or various ions and salts that may be accumulated along the way can lead to unexpected results and complications. The type of filtration system that is needed depends on the quality of water that is available to feed the system. For example, if your building provides reverse-osmosis-treated water to each lab, then the system can be limited to the filter components needed to purify and polish the water. However, if you only have utility-supplied water, you may need a reverse-osmosis unit to generate a local supply that can be used to feed the polishing system. In our opinion, simple steam distillation of water is not sufficient. In addition, the water should be filtered with a 0.22-μm tissue-culture grade filter at the last step prior to use. It is also possible to purchase water, although this would be an expensive option.

Experimental procedure
Recording Solutions

Slice dissection and recording takes place in artificial cerebrospinal fluid (ACSF) solutions (Table 1.1). These solutions consist of salts, pH buffers, energy sources, and divalent ions. Experimenters often adjust their solutions for different purposes. The replacement of sodium with NMDG or sucrose during slice preparation can improve the survival of cells in the slice (Tanaka et al. 2008), both in the brainstem and cortex. Replacing sodium ions prevents cells from spiking, reducing excitotoxic damage, and reduces the activity of Na^+/K^+ pumps, reducing metabolic demand. Sodium pyruvate and myoinositol provide alternate entry points into cellular metabolism, and their addition seems to improve cell survival. Ascorbic acid acts as a free radical scavenger and may help cell survival. Ascorbic acid may be required in experiments that use drugs sensitive to free radicals.

While the divalent ion concentrations listed in Table 1.1 are common in brain slice experiments, it is important to recognize that these are significantly higher than those occurring *in vivo*. Recent experiments in the medial nucleus of the trapezoid body at the calyx of Held have revealed how the use of these high divalent concentrations can lead to conclusions from *in vitro* studies that may not apply *in vivo* (Lorteije et al. 2009). The use of high divalents dates from the early days of slice recording, where it was found that elevated calcium levels seemed to help with forming seals between cell membranes and pipette glass. The elevated calcium concentration also increases release probability

TABLE 1.1 Composition of Recording and Dissection Solutions (in mM)

		Dissection		Recording
	Reagent	NMDG base	Sucrose base	ACSF
Salts or substitutes	NaCl	—	—	122
	NMDG	135	—	—
	Sucrose	—	240	—
	KCl	2.2	2	3
Buffers	KH$_2$PO$_4$	1.2	1	1.25
	NaHCO$_3$	20	25	25
Energy sources and antioxidants	Glucose	10	10	10
	Myoinositol	—	—	3
	Na Pyruvate	—	—	2
	Ascorbic acid	—	—	0.4
	MgSO$_4$	1.5	2	1.3
	MgCl$_2$	—	1	—
	CaCl$_2$	0.5	1	2.5
pH adjustment	HCl	135*	—	—

* Titrate pH to 7.35–7.40.

at synapses, which makes synaptic responses larger and more reliable. However, solutions with a more "physiological" calcium and magnesium concentration, such as 0.8 mM CaCl$_2$ and 1.3 mM MgSO$_4$, can ease the interpretation with respect to *in vivo* conditions. Alternatively, it is important in some experiments to provide both high release probability and low polysynaptic transmission, which can be achieved by using even higher calcium and magnesium levels (4 mM). Such concentrations are often used in photostimulation experiments where spatial maps of connectivity are the primary goal and time constraints prevent repeating maps many times to measure connections with low release probability.

When making ACSF, the components are added in the order given in Table 1.1 and are weighed out on a balance (with 0.1-mg resolution) as accurately as possible. Between uses, salts are stored in a desiccator to minimize water absorption. The divalent ions are added last, just before use, to prevent precipitation. This solution is either warmed to 34°C in temperature-controlled water baths or chilled in a freezer for 30 to 45 minutes before use. The solution should be oxygenated and pH equilibrated by gassing with 95% O$_2$–5% CO$_2$. The pH should be between 7.35 and 7.40. Inadequate gassing can lead to a more basic pH and can be a cause of poor slices. We do not recommend making "stocks" of the incubation and recording solutions for three reasons. First, without equilibration with 95% O$_2$–5% CO$_2$, the solutions will become basic over time, which causes the divalent ions to precipitate out of solution. Second, with the sugars in the solution, it does not take long to get bacterial growth, and bacterial endotoxins are not conducive to good slice health. Third, if a stock is incorrectly prepared, several days' (or even months') worth of experiments may be disrupted. With practice, it only takes about 15 minutes to prepare 1–2 L of solution, and this is readily done on the morning of each experiment.

Electrode Solutions

Pipettes are filled with an electrolytic solution whose primary function is to conduct current between the electrode and the interior of the cell (or exterior membrane surface). Because patch electrodes also facilitate the exchange of soluble molecules, pipette solutions used in intracellular recordings are designed to mimic the contents of the cytosol (or CSF, in the case of cell-attached patch) to preserve the natural function of the cell. Pipette solutions vary according to the goals of the experiment, and this is an aspect of the experimental design that requires some consideration. Standard recipes for potassium gluconate and cesium solutions are given in Table 1.2. Potassium gluconate is

TABLE 1.2 Composition of Standard Electrode Solutions (in mM). Potassium gluconate solution is commonly used for experiments performed in current clamp, while cesium solution is used for some voltage clamp experiments.

	Potassium gluconate-based solution	Cesium-based solution
K Gluconate	126	—
Cs MeSO$_4$	—	125
KCl	6*	—
CsCl	—	8*
NaCl	2*	—
HEPES (free acid)	10	10
EGTA**	0.2–5***	0.2–5***
Phosphocreatine	10	10
Na-GTP	0.3	0.3
Mg-ATP	4	4
QX-314 (Na$^+$ channel blocker)	—	3
KOH	~6****	—
CsOH	—	2****

* These values can be adjusted to manipulate the concentration of chloride. This is often done to control the reversal potential of Cl$^-$ currents. However, one needs to be careful to have enough Cl$^-$ in the solution such that it can be exchanged with the AgCl electrode; usually 5 mM is sufficient.

** Dissolve EGTA in 0.1 M KOH or CsOH first.

*** The concentration of EGTA can be adjusted depending on the amount of Ca^{2+} buffering desired. Higher concentrations may make cells more stable but lower concentrations interfere less with endogenous Ca^{2+} dynamics, which can be very important for some experiments (e.g., synaptic plasticity or calcium imaging experiments).

**** Titrate pH to 7.2.

generally used for current clamp experiments, when normal cell activity is desirable. For voltage clamp, cesium-based solutions are often used because Cs^+ ions block K^+ channels, which increases the length constant of the cell. QX-314, a Na^+ channel blocker, is often added to cesium-based solutions to block action potentials.

Electrode solutions are usually used in small quantities and so are prepared in batches and stored in single-use aliquots of 100 or 200 µL at −80°C. The usable lifetime of an electrode solution is ~3 months, although this may vary with the content. When retrieving the solutions from the freezer, it is important to vortex each solution, since during freezing components may come out of solution or a gradient in osmolarity may appear. The solution should then be centrifuged briefly to pull down any debris. Alternatively, or in addition, the solutions can be passed through a 0.22-µm filter. The solution is then stored capped in an ice bath during use and is discarded at the end of the day. The pipettes are filled as needed immediately prior to use. We have found that filling pipettes in batches that sit all day prior to use usually leads to low success rates.

The filling solutions used for patch work are often slightly hypo-osmotic with respect to the bath. There are two reasons for this choice. First, as a seal is made onto a cell, the higher osmolarity in the cell relative to the pipette helps push the cell membrane closer against the pipette tip, aiding in forming a seal with less hydrostatic pressure. Second, a portion of the cell's osmolarity is made up of elements with low diffusibility, so using a lower osmolarity in the pipette itself is less disruptive and less likely to lead to cell swelling over time.

Many electrode solutions used in patch clamp recording have unusual combinations of ions, particularly anions, that introduce electrochemical potentials both at the electrode tip and between the solution and the wire that connects to the headstage. These potentials interfere with the accurate measurement of the cell membrane potential. While the potential between the wire and the electrode solution can be eliminated by adjusting the offset controls of the amplifier easily enough, the tip potential can be problematic, and it is necessary to know its value in order to know the proper membrane potential of a cell in either current or voltage clamp. The tip potential is different when the cell is in the bath and when it has access to the cell, since the electrochemical potential in the two conditions is different. Typically the tip potential has to be measured. One way to do this is to place a filled electrode into a bath containing the electrode solution and, using an amplifier in current clamp, measure the potential. Next, while keeping a *very* slight amount of positive pressure on the electrode to minimize mixing with the bath solution, exchange the bath for a normal extracellular solution, and measure the potential again. The difference is the tip potential that would exist when the electrode is in the bath but that mostly dissipates when the electrode is in the cell. Typically, for a 140-mM K-gluconate-based electrode solution with low (4–8 mM) KCl, this potential will be about −12 mV. This means that, if the electrode potential is set to zero outside the cell and measured at −50 mV (or clamped to −50 mV) when the electrode accesses the cell, the actual resting (or holding) potential is −62 mV.

Patch Electrodes

Standard patch clamp electrodes are made from borosilicate glass pipettes that are heated and stretched to form a pipette with a blunt tip ~1–2 μm in diameter (Fig. 1.4). Since patch electrodes can only be used once and have a very limited lifespan, they are most often made onsite using programmable pipette pullers. These pullers work by heating and melting the center of a glass pipette and pulling the two halves away from each other to form two identical tapered electrodes. The exact shape of the electrode is determined by the timing and power of heating as well as pulling force. Two competing factors must be considered and balanced when pulling pipettes. First, the shape of the pipette determines its electrical resistance and capacitance, both of which should be minimized to improve recording fidelity. Typical pipettes have a resistance of 2–10 MΩ and capacitance of only a few pF. Second, pipette tips that are too large (>2.5 μm) or too small (<1 μm) may be difficult to patch with or unable to maintain long-term access to the cell.

The resistance of a patch pipette is inversely proportional to both the angle and diameter of the tip (approximated as a conical conductor). Thus, to reduce resistance, both values should be maximized; however, tip diameters greater than ~2 μm may be difficult to patch with. With the tip pulled to ~15 degrees, it should be possible to produce pipettes with low resistances around 2–5 MΩ. Such low resistance is crucial for voltage-clamp experiments measuring currents greater than a few hundred pA, such as evoked synaptic currents or currents through ion channels. For current clamp

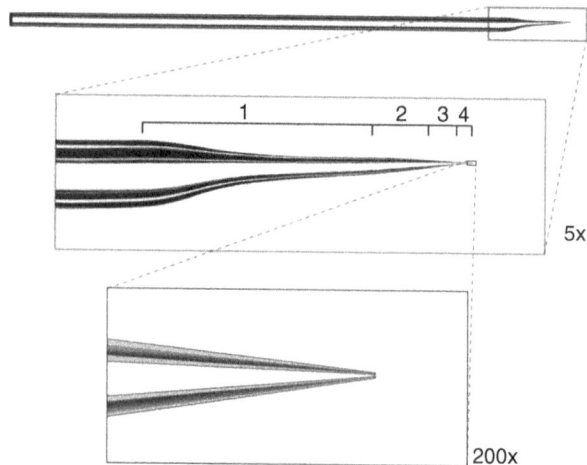

FIGURE 1.4 **Ideal patch pipette shape.** The pipette is pulled in multiple stages. The first stage is a long, narrow pull that thins the tip to help it fit under the objective. The following stages produce a rapid taper (about 15 degrees) to reduce resistance and end with a 1.5-μm tip.

recordings, higher resistances up to 20 MΩ may be acceptable as long as the access resistance to the cell is stable for the duration of the recording.

Another factor to consider is that the pipette must fit comfortably between the objective and the recording chamber (see Fig. 1.3). For objectives with a short working distance, it may be necessary to reshape the pipette tip. We shape tips by pulling in multiple stages: The first stage produces a long, narrow pull and subsequent stages taper the tip more quickly (see Fig. 1.4). The long first stage provides more room under the objective but does not significantly increase the resistance because the last 100 μm of the tip accounts for roughly 95% of the total resistance.

The choice of glass can be important in some experiments, and so familiarity with the different compositions made by different manufacturers can be helpful. Typical pipette glass is borosilicate-based and will have an outside diameter of 1.0–2.0 mm and an inside diameter of 0.5–1.6 mm. The thickness of the glass wall is generally maintained in proportion to the diameter of the tip as the glass is pulled, and for patch clamp recording, thicker-walled glass is generally preferable to reduce pipette capacitance. Patch pipette glass often includes a small interior filament that acts as a wick to draw electrode solution into the tip. Some pipettes are produced with the raw ends fire-polished. This is necessary to prevent the otherwise sharp glass from scratching the thin AgCl layer on the silver wire. Pipettes can be easily polished by holding the back end over a small flame for a few seconds, until the glass glows orange.

Producing clean, correctly shaped pipettes often requires much trial and error. Once pipettes are pulled, they may be individually inspected under 10× and 40× objectives. This screening process is used both to guide adjustments to the puller to obtain the desired diameters and tip shapes, and to discard pipettes that are broken or fouled or that fall outside the desired tip diameter. It is of the utmost importance that the tips of patch electrodes be clean. Thus, pipette blanks should only be handled by the ends to avoid placing skin oils in the region that will be heated. Pulled pipettes are only used on the day they are made, because of increased chances of tip fouling and potential hydration of the fine tip glass that could affect the dielectric properties of the glass and introduce recording noise.

An optional final step in preparing pipettes is to coat the tip to reduce capacitance to the bath and improve the dielectric properties. The effect of the coating is to reduce recording noise and to improve the ability to fully and properly compensate the electrode capacitance for voltage clamp recordings. When performing voltage clamp studies, we consider the application of a coating essential. In experiments in which only current clamp recordings are done, this step can be skipped, although the reduction of capacitance reduces the amount of compensation needed, which in turn reduces the overall noise level of the recording. There are two approaches that are commonly used. The first is to use a conformal coating, such as Sylgard (Dow Corning 184). This is a two-part mixture that can be painted to within 50 μm of the tip using a fine needle while viewing the tapered region of the pipette with a dissection microscope. The mixture cures in several seconds by applying heat (we use a paint stripper on its low setting). The Sylgard can also be stored uncured in the freezer (−20°C) for about 2 weeks, and we find that preparing the mixture about 24 hours prior to first use is also helpful. A second approach is to wrap the tip of the pipette with a 2- to 3-mm-wide strip of Parafilm, which can be melted with gentle heat, or to dip the tip of the pipette in molten Parafilm, while keeping positive pressure on the pipette to maintain a clear tip.

Filling the pipette with electrode solution is also an important step and one where problems can occur. While there are various commercial filling needles, these tend to be expensive and hard to keep clean, often leading to clogged electrodes. A different approach is to use 100-µL pipette tips (those made with a harder plastic seem to work best) pulled over a flame to create a very fine tube (Fig. 1.5A, B). With care and practice, these make

(a)

(c)

(b)

(d)

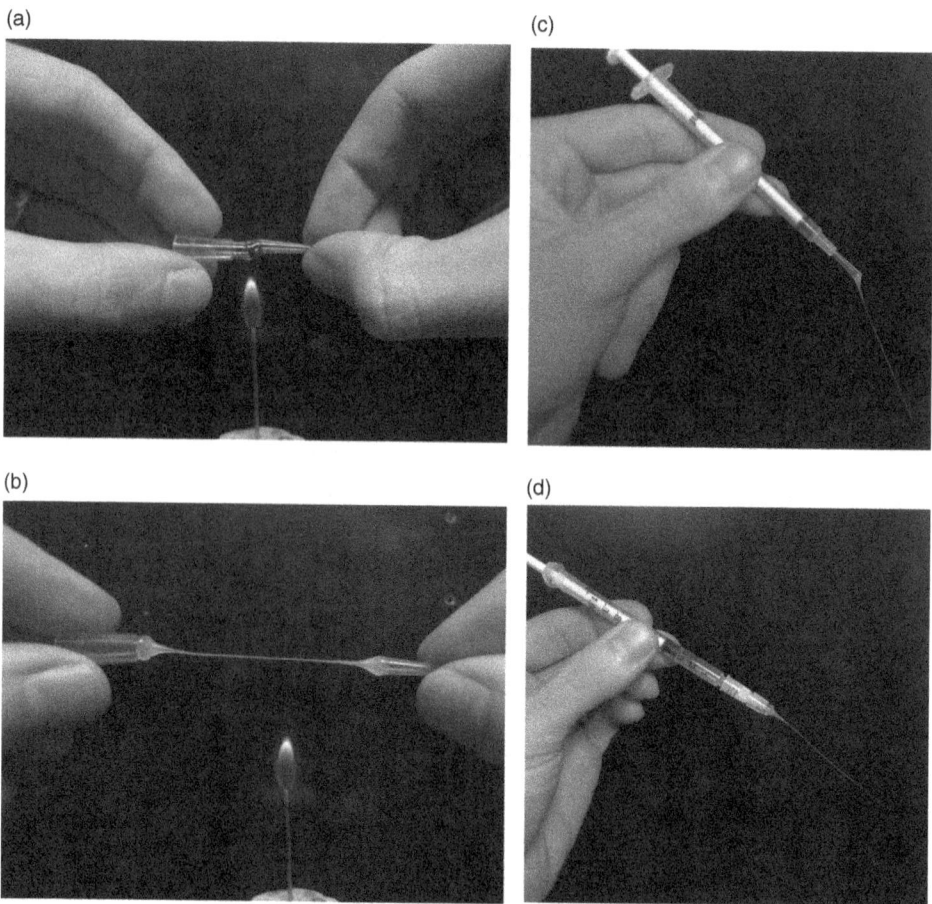

FIGURE 1.5 **Making patch pipette fillers from disposable pipette tips.** A: Heating a 100-µm disposable pipette tip over a small flame. The tip should be rotated to produce even heating and care should be taken to avoid burning the plastic. B: As soon as the tip has melted through, remove it from the heat and pull into a thin tube (this takes some practice). Cut the tube with a sharp blade to avoid crushing it. C: Filler made from tip of pipette inserted into 1-mL syringe. D: Filler made from base of pipette attached to 1-mL syringe and a low-volume, 0.2-µm-pore syringe filter.

filling tips that are only a few hundred μm in diameter. The fillers can be inserted into a 1-mL syringe and backfilled from the electrode solution stock (see Fig. 1.5C) or capped over a prefilled syringe and filter (see Fig. 1.5D). The major advantages of these fillers are that they are disposable if they become dirty, are economical, and are easily remade. For making fillers and fire-polishing pipettes, we have created a small burner from an 18-gauge blunt needle that provides a flame size similar to a match.

Dissection and Slicing

The preparation of brain slices containing healthy cells is critical to the success of patch recording. The goal is to extract a section of the brain such that the cells of interest are close to the surface of the slice and any other required network connections are intact elsewhere in the slice. Furthermore, we need to make sure that the cells or tissues are still sufficiently alive and undamaged and that they can be visualized well enough to facilitate patching. Producing viable brain slices can be very difficult, and proven methods often vary widely between brain regions. The main factors affecting slice viability are as follows: (1) Prevention of ischemic damage by dissecting and slicing quickly, often in well-oxygenated, ice-cold ACSF (the ACSF may actually be partially frozen); (2) Prevention of excitotoxic damage through use of specialized ACSF solutions; and (3) Prevention of mechanical damage by avoiding compression or stretching of brain tissue and by using well-tuned slicers with appropriate blades.

Blades. The choice of cutting blade can be critical to successful slice preparation, especially in older tissue. The most commonly used blades are commercially available double-edged stainless-steel razor blades. These vary in quality, however, and different types should be tried to determine which ones work best for a specific brain region. "Platinum plus" blades have worked well in the brainstem and cortex, while other types of blades have been found to yield very poor cutting. Reusable blades made of sapphire or ceramic are also excellent choices, especially if they can be resharpened. Blades should be cleaned prior to use, and stainless-steel blades should only be used for one cutting session. Cleaning is necessary to remove oils and other protective chemicals used to retard oxidation and corrosion of the blades. We clean by first briefly washing the blade in acetone with a cotton swab, followed by a 70% ethanol rinse, and finally a distilled water rinse. The blade is then dried and placed in the chuck of the slicer.

We discuss preparation of slices from two different brain regions below to illustrate two different approaches to creating viable brain slices. The first method, for neocortex, follows a conventional approach, while the second method, which we use for cochlear nucleus, demonstrates how variations on the procedure may be best applied for different brain regions. Prior to removing the brain, all solutions need to be at an appropriate temperature and properly oxygenated, surgical tools should be located and clean, and the cutting blade should be in place. The goal is to minimize the time between decapitation and the incubation of the slices in the holding chamber. At the same time, it is critical to be careful with the tissue and to handle it gently. In each approach, the animals are first deeply anesthetized, according to an approved protocol, and decapitated.

Cortex. The skull is exposed, cut down the midline with fine-tipped scissors, and peeled back with rongeurs (in adult animals where the skull is thick) or with fine-tipped scissors (in younger animals when the skull is thin). Care should be taken not to touch the brain itself when removing the skull. The brain is removed after carefully cutting major cranial nerves that may enter or leave near the tissue of interest. The brain is then "rolled" out using a small spatula into an ice-cold dissection solution (see below for composition). The tissue is trimmed, using fine scissors and scalpel blades. The key elements in trimming are to obtain a flat surface that is parallel to the desired plane of section that can be used to glue the brain block to the stage, and to remove any excess tissue that does not contribute to stabilizing the brain block during cutting. It is helpful to have a specific sequence in which the trimming is performed, as with practice this can greatly speed the preparation.

The next step is to place the tissue block on the chuck that will go into the slicer. We usually prepare the mounting position by laying down a small platform made of 4% agar (made in 150-mM NaCl) to support the tissue, and place an agar wall behind the platform. In some cases, these agar supports are cut with an angled surface to help orient the brain when it is glued down. A drop of cyanoacrylate glue is placed on the platform just in front of the wall. The tissue is then picked up by sliding it onto a small piece of ashless #50 filter paper (this paper is "hard" and can hold small blocks of tissue even when wet), such that the part of the tissue that will be against the wall is against the tissue paper. The chuck is placed at an angle such that the wall can support the tissue by gravity. The tissue block is then transferred onto the cutting chuck by sliding it against the wall until it comes in contact with the glue, at which point the filter paper is slid out from under the tissue. It is important in this step that the glue not come in contact with the filter paper. We next mount 4% agar support blocks (usually ~2 × 2 × 6-mm posts) that are glued to the stage and gently abut the tissue, to minimize movement of the tissue during cutting. Cutting takes place in a cold solution in a previously frozen cutting chamber surrounded by an ice slurry.

Brainstem (cochlear nucleus). The squamous portion of the occipital bone over the cerebellum is removed with rongeurs (fine scissors are sufficient in mice), exposing the cerebellum and brainstem. The brainstem is briefly washed with warmed oxygenated ACSF. The temporal bone is carefully retracted laterally, the floccular and parafloccular lobes of the cerebellum are gently lifted, and the exposed eighth nerve (both the auditory and vestibular branches) is then sectioned with the tip of a #11 scalpel blade. Care is taken to minimize stretch of the nerve while cutting. The brainstem is transected rostral to the inferior colliculus with a spatula, and again caudal to the obex, removed from the skull, and rinsed again in ACSF. A small tissue block containing the cochlear nucleus of the left side is then isolated from the brainstem. The brainstem is bisected at the midline longitudinally with a scalpel, and trimmed rostral and caudal to the cochlear nuclei with scissors. The choroid plexus lying above the cochlear nucleus is gently teased away with #5 forceps. Most of the cerebellum is cut away at the cerebellar peduncles with scissors. The rostral and caudal ends of the block are trimmed at an angle approximately parallel to the long axis of the cochlear nucleus. A final cut is made parallel to the desired cutting plane; this may be along the midline or across the ventral surface. This block is transferred using a strip of hardened #50 ashless filter paper, blotted to remove excess fluid, and mounted on the chuck of the tissue slicer with cyanoacrylate glue. The tissue is supported with agar blocks from behind and on the sides. The chuck and tissue are immersed in a *warmed* carbonated cutting solution in the bath.

For either tissue region, slices are cut by carefully advancing the blade into the tissue under visual control. Often, a few 500-μm-thick slices are quickly taken until the desired region is reached, and then the cutting thickness is adjusted to 250–350 μm, and a series of slices are collected. As each slice is taken, it is checked under a dissecting microscope to be sure that it is from the appropriate region and is not damaged, and then it is transferred to the incubation chamber using the blunt end of a Pasteur pipette whose tip has been broken off and fire-polished so a pipette bulb can be attached. We prefer this method over using small paintbrushes, as there is less mechanical stress to the tissue slice during the transfer.

Slice Incubation

Slices are commonly allowed to incubate for 30–60 minutes at 32–34°C after slicing to allow them time to recover and re-equilibrate to the ACSF environment. After this the slices are usually incubated at room temperature. The slices are held in an incubation chamber that may be in a water bath, or on the bench for room temperature incubation. We primarily use a simple chamber that consists of a 100-mL glass beaker. A sintered-glass gas dispersion tube is inserted into the chamber and ~60 mL of ACSF is added, and well gassed with 95% O_2–5% CO_2. The slices rest on a permeable nylon mesh that forms the bottom of multiwell tray. This tray is hung in the chamber so that the ACSF covers the mesh but is 1–2 mm from the top of the tray walls. The top of the chamber itself is loosely covered with Parafilm to minimize evaporation over time. Other incubation chambers have been used as well. The key requirements are that bathing solution is exchanged around the slices by a stirring action (usually provided by the gas dispersion system), that the solution is well gassed, that evaporation is controlled, and that the chamber and gas dispersion system do not leach chemicals into the incubation solution.

Recording Chamber, Perfusion, and Harps

After incubation, slices are held in a small-volume (~0.3 mL) recording chamber mounted on the stage of an upright microscope. The slice rests either on a coverglass or on a small section of netting, and is held in place by another net stretched across a stainless steel harp. Warm, oxygenated ACSF is perfused over the slice at 2–8 mL/min by siphoning from a flask. The incoming perfusate is warmed to 38°C just prior to entering the chamber with a feedback-controlled heater, resulting in a solution temperature at the slice of $33 \pm 1°C$. While some experimental procedures, such as analysis of network organization using photostimulation, can be done at room temperature, it is nearly always preferable to record at elevated temperatures to more closely approximate the normal kinetics of ion channels, the release properties of synapses, and the engagement of intracellular signaling cascades.

The fluid level in the recording chamber is regulated by positioning an aspirator above the surface of the water. While a properly configured aspirator should maintain the bath at a constant level, it is nevertheless important to monitor the chamber to avoid overflows, which may damage equipment, or underflows, which may damage the slice.

To prevent electrical noise, it is very important that the fluid in the recording chamber be electrically isolated from everything except the patch and ground electrodes. If the chamber overflows, this might create a new electrical path to ground or other parts of the setup, introducing a new potential noise source.

Patching

Finding viable cells. You are finally ready to patch a cell. The first task, then, is to find a cell that appears to be healthy and is in the correct location of the slice. The appearance of healthy cells will vary somewhat between brain regions, but typically these will appear to have a smooth, translucent interior and a smooth cell membrane (Fig. 1.6, *white arrows*).

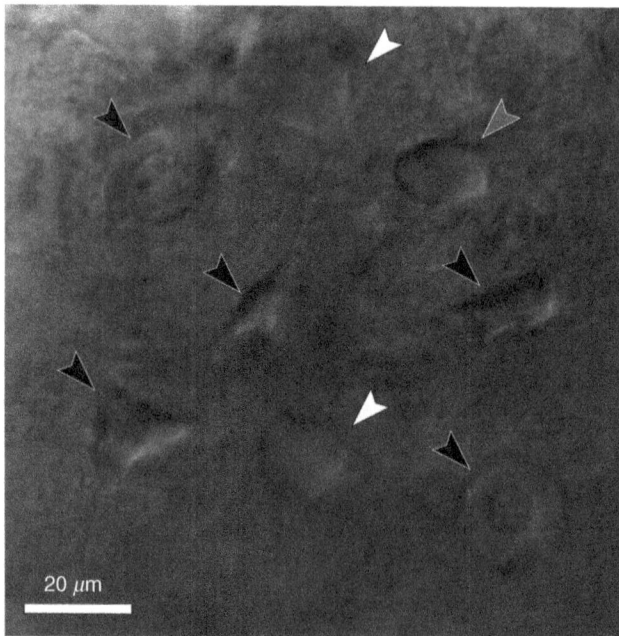

FIGURE 1.6 Neuron examples in a cortical brain slice under gradient illumination. *Black arrowheads* indicate unhealthy or dead cells, *white arrowheads* indicate healthy cells, and a *gray arrowhead* indicates a borderline cell. (This figure is a composite of multiple images from different regions of a slice.)

Unhealthy cells often appear either shriveled or bloated, have a rough or abnormally transparent interior, or have a visible nucleus and nucleolus (Fig. 1.6, *black arrows*). However, these features are not always diagnostic of cell health, and it is possible to bias the cell selection by avoiding healthy cells with an abnormal appearance. Ultimately, trial and error may be the best way to determine which cells can be patched successfully and which appearances are associated with healthy cells. In a few cases, the situation may call for alternative methods such as shadow patching (using fluorescent electrode solution and looking for dark regions indicating a healthy cell that is impermeable to fluorophore) or blind patching (using electrical signals rather than visualization to determine when the pipette has contacted a cell) instead of direct visualization.

It is preferable to avoid cells very close to the slice surface, as they are likely to have been severely damaged during slicing. Generally, one should attempt to patch cells as deep as possible given the limitation of visibility in the slice. For relatively transparent tissue such as neocortex from young animals, it may be possible to visualize cells up to 50–100 μm deep. For older or heavily myelinated tissue, visibility may be limited to 20 μm or less. In this situation, having properly adjusted illumination and a good camera is crucial. It may also be necessary to use software that allows large contrast adjustments or background subtraction.

Once a candidate cell has been selected, center the cell within the camera's visible range and switch back to a low-power objective in preparation for positioning the patch pipette.

Filling patch pipettes. For each cell you wish to record, a new patch pipette must be prepared and filled with electrode solution. It is recommended to make all patch pipettes at the beginning of the day (about 6–12 should suffice, depending on the experiment), but do not fill the pipettes with electrode solution until immediately before each is to be used. The day's aliquot of electrode solution should be thawed, vortexed, centrifuged, and kept chilled on ice. Note that vortexing is critical because just-thawed electrode solutions may have a large osmolarity gradient from the top of the tube to the bottom.

To fill the pipette, either (1) attach a plastic filler (see Fig. 1.5) to a 1-mL syringe and draw a small amount of electrode solution into the syringe, or (2) draw all of the electrode solution into the syringe and cap with a 1-mL aliquot filter and a plastic filler (see Fig. 1.5C, D). Insert the filler as far as possible into the pipette and inject enough solution to fill approximately two thirds of the pipette. The exact amount needed will be just enough such that the electrode solution makes contact with the AgCl wire in the pipette holder. If the pipette is overfilled, electrode solution may seep into the pipette holder, and this can add noise to the recording. Air bubbles between the AgCl wire and the tip of the pipette may increase the resistance of the electrode. These can be removed by sharply tapping the pipette against the counter.

Many modern pipette blanks have a glass filament that is fused to the inside of the tubing. The filament creates a capillary flow of fluid that helps bring the electrode solution to the pipette tip. It can also act as an electrical conductor. These pipettes can be filled from the back and the tip will fill over several seconds to a minute, with no bubbles in the tip. In the absence of such a filament, pipettes can be also filled from the tip by applying suction to the back while the tip is dipped into the recording solution. This will bring a tiny amount of the solution into the tip, and then the pipette must be filled the rest of the way from the back.

When the pipette is filled, place it over the AgCl wire and secure it snugly into the pipette holder. Do not overtighten the retaining cap, as this can put torque on the rubber sealing gasket and cause the pipette tip to drift slowly as the rubber relaxes.

Approach. The most important thing to remember while patching is that clean glass is very sticky. Whatever touches the tip of your pipette first will adhere to it permanently. We keep positive pressure inside the pipette so that electrode solution is constantly flowing through the tip, ensuring that nothing touches it until we are ready. Most electrophysiologists use a syringe to provide this pressure, while some prefer to use their mouth. It may also be helpful to use a pressure gauge.

To begin, you should be looking at the slice through a low-power objective that is still centered on your target cell. Using the micromanipulator, position the tip of the pipette in the center of the field of view above the fluid surface. Make sure there is positive pressure on the pipette, then lower it into the fluid but do not yet touch the brain slice. Note that sometimes salts may crystallize or debris may appear on the surface of the bathing solution; these should be aspirated from the surface, or the cause identified and eliminated, because they can foul the tip of the pipette as it enters the solution, making patching difficult or impossible.

At this point, adjust the amplifier's pipette offset (this is described in the manual for your amplifier). Configure your amplifier to output 20-ms, −10-mV pulses (or for current clamp, use 1 nA) and calculate the resistance of your pipette from the recorded response (Fig. 1.7). Most electrophysiology software will have built-in features for measuring pipette resistance. Monitor the pipette resistance continuously until the cell is patched. If the resistance increases unexpectedly, discard the pipette (it has probably clogged). Record the pipette resistance for every patched cell as this information can be very useful when analyzing data.

Into tissue. With the pipette tip still above the surface of the slice, switch to a high-power objective. If the pipette tip is no longer visible, you may need to move the tip a small distance until it appears in the view. Be cautious—if you cannot see the pipette, it is easy to accidentally drive it into the slice or the objective. Once the tip is visible, proceed slowly down toward the surface of the slice, continuously refocusing to keep the tip in view.

For most situations, it is acceptable to simply center the pipette tip over the position of the target cell and descend directly downward to the cell. For tougher, myelinated tissue or for deeper cells, it may be advantageous to push the pipette in diagonally along its axis. This will avoid some amount of tissue compression that may result from going straight downward. Most micromanipulators can be configured for this purpose.

(a)

(b)

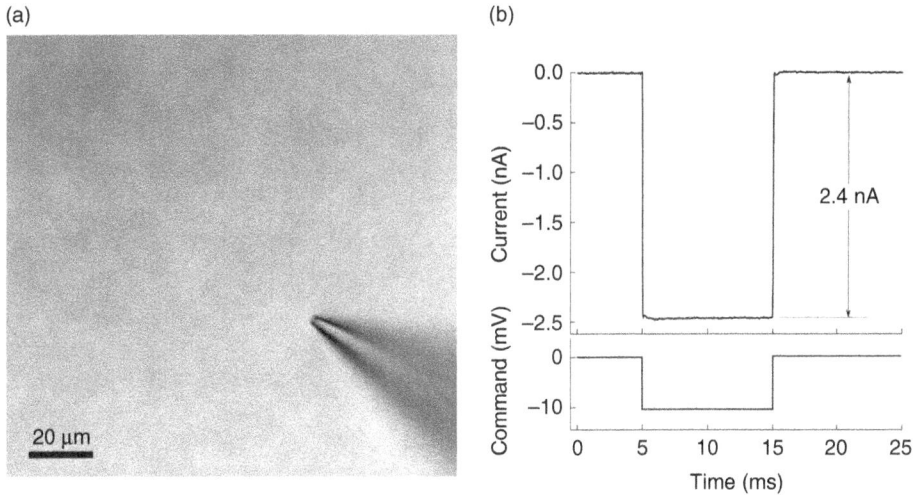

FIGURE 1.7 **Voltage clamp recording from a pipette in the recording chamber bath.** A: Photo of patch pipette in bath far above the slice. B: Voltage clamp command and current recording from patch pipette in bath. The voltage clamp requires 2.4 nA of current to effect a 10 mV pulse, indicating a pipette resistance of 4.2 MΩ.

As the pipette tip enters the surface of the slice, you should immediately see the tissue gently spread away due to the pressure in the pipette. If you do not see this, then it is likely the tip is already clogged or there is insufficient pressure (and this pipette should be immediately discarded). Too much spread, however, is not a good sign, as it means that you are flooding the slice with the electrode solution. Depending on the solution, this can depolarize nearby cells, and the mechanical action of the flow may also disrupt the tissue. If this occurs, reduce the pressure.

As you proceed closer to the cell, you may encounter obstacles such as fibers or other cells. The positive pressure will push some obstacles out of your way, while other obstacles will need to be avoided. Some trial and error may be necessary at first. Remember, the goal is to arrive at the target cell with a clean pipette tip. Positive pressure makes this possible, but some situations will require finesse as well.

Near cell. When the pipette tip is within 10 μm or so of the target cell, correct the amplifier's pipette offset again. Press the tip slowly into the center of the cell to form a visible dimple (Figs. 1.8A and 1.9). This dimple is your indication that there is indeed nothing else between the pipette and your cell. Release the pressure on the pipette and wait while monitoring the resistance of the electrode. The cell membrane should immediately come into contact with the electrode tip and begin to form a seal.

The membrane that has adhered to the pipette tip may spontaneously rupture, so it is important to prepare for this. After the seal resistance has increased past ~100 MΩ, voltage clamp the pipette near the estimated resting potential of the cell, less the junction potential (e.g., if a typical cell rests at −75 mV and the junction potential is −12 mV, voltage clamp the pipette at −63 mV with brief steps to −73 mV). The resistance should

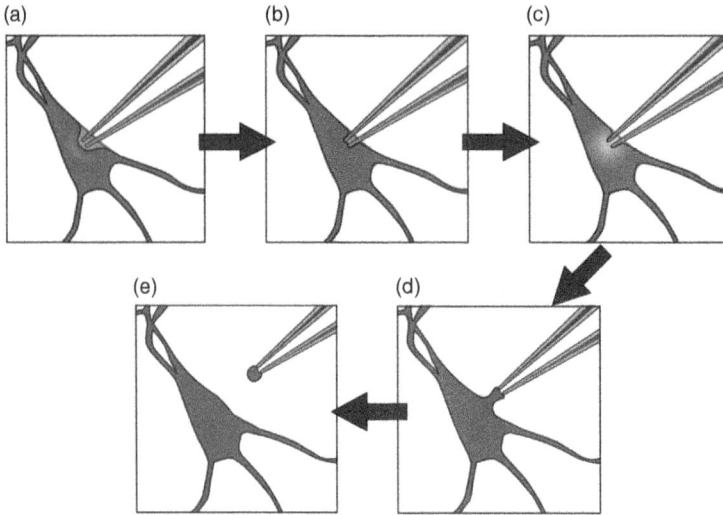

FIGURE 1.8 Patch procedure. A: Approach the cell with positive pressure in the pipette. The surface of the cell should form a visible dimple. B: Release pressure on pipette, and then apply gentle suction to seal the membrane against the pipette. This is the *cell-attached* configuration. C: Apply sharp suction to the pipette to rupture the membrane, granting electrical access to the cell interior. This is the *whole-cell* configuration. D: From the whole-cell configuration, pull the pipette very gently away from the cell until (E) the membrane separates and recloses. This is the *outside-out* configuration.

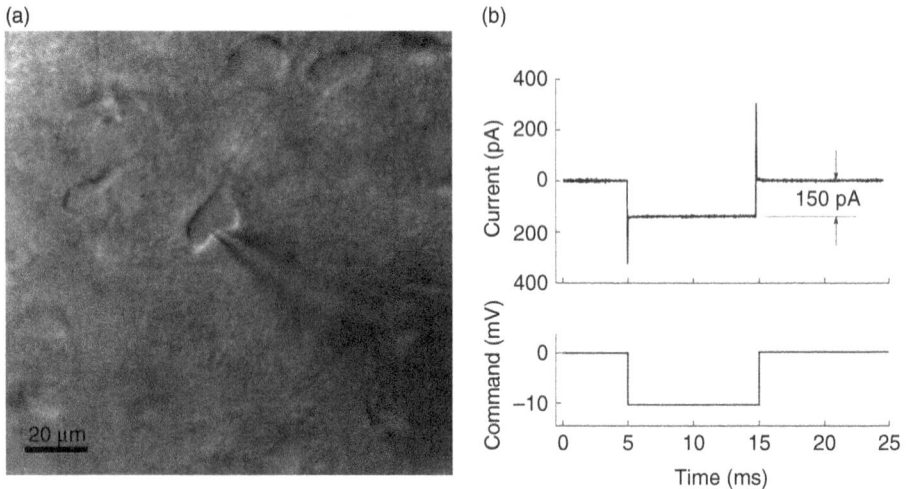

FIGURE 1.9 Forming a seal on a cell. A: A "dimpled" cell immediately before being patched. B: Voltage clamp recording shortly after releasing pipette pressure. The resistance at the pipette tip has increased to 66 MΩ.

continue to increase over a few seconds to a minute, going from a few MΩ to over 1 GΩ (Fig. 1.10). If resistance is not increasing quickly enough, gentle suction on the pipette can encourage a seal to form. If your software allows, it can be very helpful to watch the pipette resistance plotted over time.

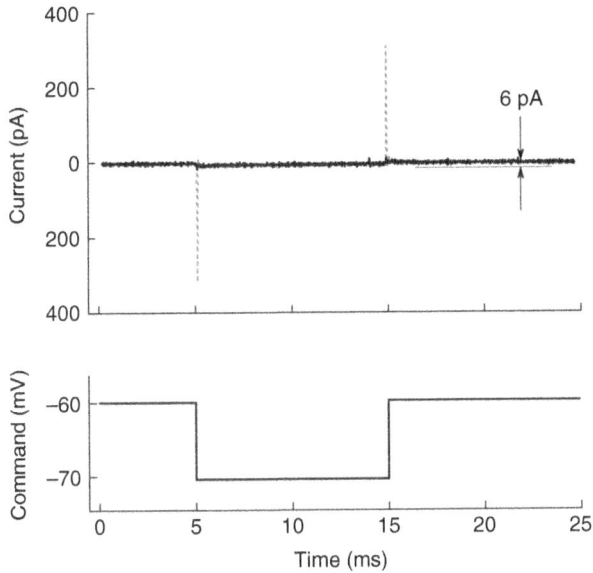

FIGURE 1.10 Voltage clamp recording from cell-attached pipette before (*dashed line*) and after (*solid line*) adjusting the pipette capacitance compensation. The seal resistance has increased to 1.6 GΩ.

After forming a gigaohm seal, the pipette is considered "cell attached" (see Fig. 1.8B). In this mode, it is possible to cleanly record action potentials from the patched cell, but little else should be visible. If your amplifier has built-in pipette capacitance compensation, now is the perfect time to adjust those settings (your amplifier manual should discuss this in detail). This will minimize the transients at the beginning and end of the voltage command step (see Fig. 1.10).

Break-in. Once a gigaohm seal has been formed, access to the cell can be obtained. Apply brief pulses of suction to break the membrane within the lumen of the pipette (see Fig. 1.8C). There are several ways to do this. In our lab, we typically use a 1-cc tuberculin syringe to create the suction, using small, quick pulls on the plunger (0.01- to 0.03-cc displacements). The negative pressure needed to break into the cell varies with cell type, the preparation, and the pipette tip diameter and taper. Under the best conditions, a displacement of <0.1 cc is sufficient to provide a clean break-in. Once the break-in is achieved, the negative pressure is immediately released. A traditional way to apply suction is to use a mouth-pipette tube. This also gives good control of the pressure. Another way is to use a controlled negative-pressure-generating system, such as a column of water, along with a valve. However, the complexity of such a system may not be worth the effort to maintain it compared to using the simpler methods. Many amplifiers also offer "zap" and current pulse controls that can be used to try to break the luminal membrane by voltage breakdown. However, we have not found these to be very effective in the cochlear nucleus or auditory cortex, and when access to the cell is achieved it has high resistance and is not stable.

When whole-cell access is obtained, the membrane current trace will consist of a fast transient current that decays back toward the baseline (Fig. 1.11). The amplitude of this transient is inversely proportional to the series resistance (lower resistances generate larger transients), while the time constant of the transient decay is approximately the product of the series resistance and the effective cell capacitance seen by the electrode. The access resistance should be low. Using –10-mV steps, a –1-nA peak current would correspond to 10-MΩ access and –2 nA to 5 MΩ (as follows from Ohm's law). The time constant corresponds to the speed of the uncompensated voltage clamp, and the fast component, corresponding to charging of the soma, usually should be well under 1 ms.

Compensating series resistance and whole-cell capacitance. In an ideal voltage clamp, the membrane potential at the patched cell is exactly equal to the requested command voltage. In practice, several factors prevent perfect control of the membrane potential. The small tip of the electrode, combined with cell debris inside the tip, creates a series resistance between the interior of the electrode and the interior of the cell. When current is passed through this series resistance, it results in a voltage difference, so that the resulting membrane potential is no longer equal to the command voltage but is shifted in a direction that depends on the sign of the current flowing through the electrode at any instant in time. When currents are large, the resistance of the bath electrode may also contribute to this error. Series resistance, when combined with the capacitance of the pipette and of the cell, also introduces a complex low-pass filter that affects how

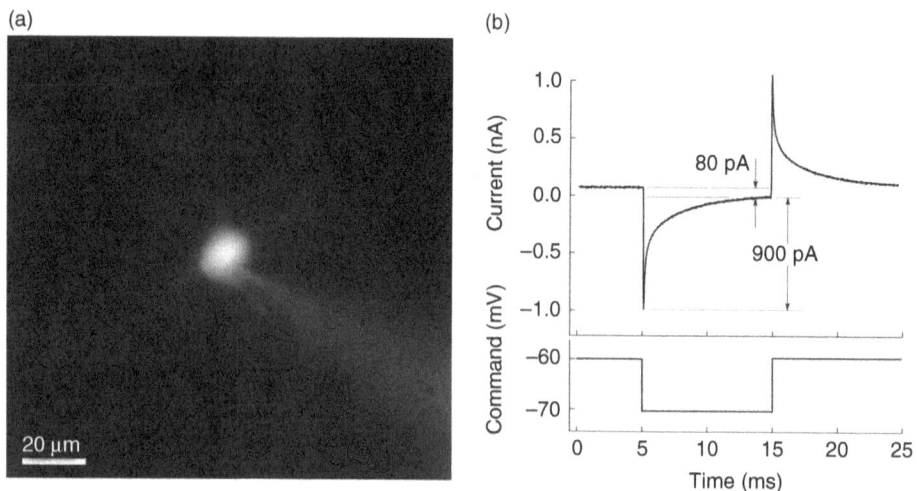

FIGURE 1.11 Whole cell recording mode. A: Whole-cell patched neuron filled with fluorescent dye. B: Voltage clamp recording from the same neuron. The steady-state current is ~80 pA, indicating an input resistance of 125 MΩ. The peak of the charging transient is 900 pA past the steady state, indicating an access resistance of 11 MΩ.

rapidly the voltage at the cell can be changed and how rapidly the amplifier can detect changes in the cell voltage.

Voltage clamp amplifiers include compensation circuitry that attempts to correct these effects by including a feedback circuit that takes into account the series resistance and cell capacitance, and injects current through the electrode in an attempt to faithfully follow the command voltage. This compensation is essential for experiments that require precise control of the membrane potential and accurate recordings of fast or large currents. It also effectively increases the bandwidth of the clamp, resulting in tighter control of membrane potential during rapid changes in membrane conductance.

The drawbacks to series resistance compensation are that it introduces additional high-frequency noise to the recording, and it is prone to producing oscillations that may damage or destroy the cell if configured incorrectly. For recordings that require very low noise, where the currents are slow, and where a voltage error can be tolerated or is demonstrably small (e.g., measuring small currents where the voltage error is also small), it may be preferable to disable series resistance compensation.

The accuracy of compensation is limited by the extent to which the user can adjust the settings to closely reflect the electrical circuit of the pipette and cell. The details for configuring series resistance compensation are found in the manuals of the amplifiers, and because the compensation circuitry varies between amplifiers, those recommendations should be followed. A few important points are in order, however. First, any cell with an extended dendritic tree will have a capacitive transient that has multiple time constants. However, the amplifiers are all designed to compensate a single time constant (e.g., a spherical cell body with no processes). Thus, care in adjustment must be used to focus on the correct (somatic) time constant. Second, stable recording conditions need to be attained. Any change in access resistance, or even the bath fluid level, can affect the conditions needed for optimal compensation and will result at best in incorrect compensation and at worst in the system going into oscillation and destroying the cell.

At this point, we wish to raise an important limitation of voltage clamp that is all too often ignored in the literature. Only the point of the cell immediately adjacent to the electrode is properly "voltage clamped." There is a large and local spatial gradient over which clamp fidelity decreases, usually on the order of 100 μm or so (Fig. 1.12). This can be partially corrected by using the amplifier's compensation circuitry and by using electrode solution that blocks ion channels to increase the electrotonic space constant of the cell. This issue has been discussed by a number of authors over the years (Spruston et al. 1993, Williams and Mitchell 2008). In most central neurons, even under the best conditions, only the cell body and proximal ~50 μm of the dendrite are under good control of the voltage clamp. Thus, it may be advisable to record in current clamp to minimize the potential for voltage clamp problems—or if appropriate, patch recordings directly from dendrites might be considered.

Even though the entire neuron cannot be clamped, recording in a voltage clamp mode has several advantages for examining synaptic responses. The clamp keeps the

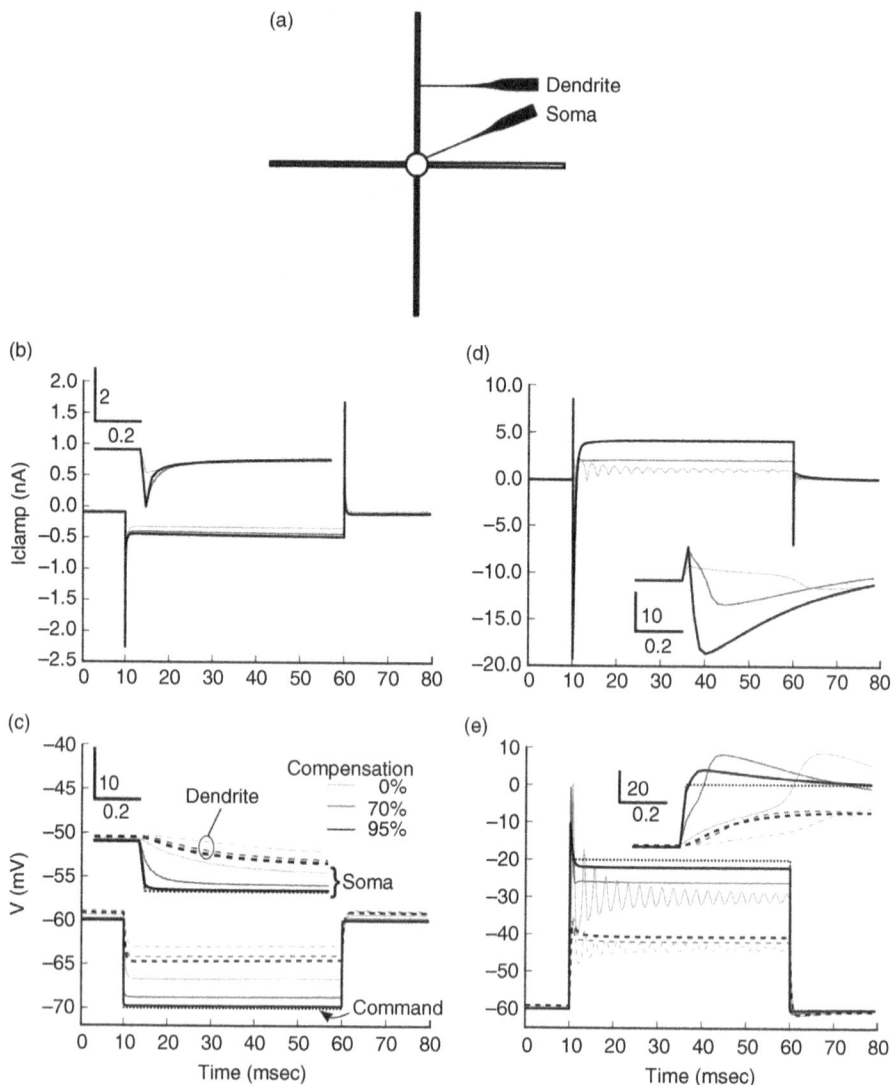

FIGURE 1.12 The effects of pipette capacitance and series resistance compensation (0%, 70%, and 95%) on a simulated neuron (soma, 20 μm in diameter, dendrites 200 μm long and 1.5 μm in diameter). A: Schematic of neuron; a single-compartment soma with four dendrites. The somatic electrode is used to voltage clamp the cell, while the dendritic electrode records voltage (in current clamp) at a mid-dendritic position 100 μm from the soma to illustrate the inadequate space clamp. B: Clamp currents (somatic electrode) in whole-cell recording mode, for three different levels of compensation for voltage steps from −60 to −70 mV. Compensation increases the amplitude of the capacitive charging transient, as well as the steady-state current. Inset: Initial transient on an expanded time scale. C: Comparison of voltage at the soma for different amounts of compensation, compared to the command voltage (*short dashed lines*). *Longer dashed lines* indicate voltage recorded at the mid-dendrite position. Note that until compensation approaches 95%, the somatic voltage differs from the command voltage. The dendritic voltage control is not improved by increasing compensation, and at best the voltage step is only about half of the command step. D: Currents in response to depolarization from −60 to −20 mV. With no compensation, the sodium conductance leads to an oscillatory current, corresponding to unclamped partial action potentials. Increasing compensation brings the currents under better control (inset) and also increases the outward potassium current. E: Command voltage and actual voltage at the soma and dendrites. Note the action-potential–like waveforms with no compensation. Even with 95% compensation the voltage in the dendrite (*long dashed lines*) varies with time and does not reach the command level. B–E: Insets show the first 1 msec of the current traces (B, D) or voltage (C, E) for each compensation level. Model: "Type I" neuron (Rothman and Manis 2003), with sodium, delayed rectifier, hyperpolarization-activated cation conductance, and leak conductance, with $R_i = 150\ \Omega$-cm.

membrane potential relatively constant and below spike threshold, so that synaptic inputs are not likely to drive spikes unexpectedly. In addition, voltage clamp can largely remove the effect of membrane capacitance from conductance changes generated near the recording electrode, which can improve the signal-to-noise ratio for detection and measurements of synaptic inputs, especially when measuring single quantal events such as miniature excitatory or inhibitory postsynaptic potentials. Finally, even though dendritic synaptic events are not fully clamped, relative changes can be measured under different conditions in individual cells, while holding the membrane potential (and synaptic driving force) constant. However, this requires careful consideration of the potential influence of any manipulations on the quality of the clamp and on nonuniform changes in driving force across the dendritic tree.

Inside-out, outside-out. Outside-out patches (see Fig. 1.8D, E) are very useful in evaluating the voltage dependence of ion channels and the kinetics of neurotransmitter receptors. These patches have a very low capacitance and can be well controlled under voltage clamp. They can be pulled from cells in slices, including from fine dendrites. A major advantage of using isolated membrane patches is that the space clamp problem is eliminated. A second advantage is that the site where the channels with particular currents are located can be determined. Disadvantages include the fact that the channel function may be disturbed by washout of essential proteins or intracellular ions, or even by introducing an unusual curvature to the patch membrane.

To pull an outside-out patch, first obtain a whole-cell recording (see Fig. 1.8C). We find that the best patches are pulled within about 10 minutes of accessing the whole-cell configuration, and this provides time to perhaps fill the cell with a dye and to obtain a characterization of the intrinsic physiology. Next, switch the amplifier to voltage clamp, holding the cell at −60 mV, and provide 10- to 50-msec-long voltage pulses going to −70 mV. It is best not to compensate the amplifier at this point, as visualization of the access resistance and clamp time constant is better obtained by watching the uncompensated currents on an oscilloscope. Begin forming the patch by slowly drawing the pipette away from the cell, at a rate of a few μm per 10 seconds, stopping frequently, in a direction normal to the cell surface at the point of contact (see Fig. 1.8D). Watching the oscilloscope, you should see an increase in access resistance, as indicated by a decrease in the peak current at the beginning and end of the step, once the pipette is more than ~5 μm from the cell, and a thin bridge of membrane may be visible connecting the pipette to the cell. Continue pulling slowly until the capacitive charging transient at the beginning and end of the voltage pulse becomes very small, and the input resistance of the patch increases as indicated by a decrease in the small steady-state current during the step. The holding current should be less than a few tens of pA. You should be able to continue pulling the pipette away and up so that the tip is in the bath above the slice (see Fig. 1.8E). At this point you have an outside-out patch. Applying voltage steps may reveal small currents (10–200 pA), especially with depolarization. Loss of the patch is indicated by a large increase in the holding current and noisy traces.

Note that this procedure can also result in a resealing of the membrane of the cell that the patch was pulled from, and it is often possible to image the cell after recordings from the patch are complete.

Running the experiment. When the cell is patched, you are ready to run your experiment. Patched cells can be temperamental, so it is important to monitor the health of the cell for the duration of the experiment. The major indicators of a failing cell are decreased or increased resting membrane potential (10–15 mV above or below the typical resting potential) and decreased input resistance. These can be monitored by periodically recording the response to current or voltage pulses similar to the procedure used during patching. Alternatively, the cell can be continuously monitored by generating an audible signal from the amplifier.

Additionally, access resistance may increase during the experiment. Although some increase in access is normal, it may cause problems if it continues to increase past 15–20 $M\Omega$ in current clamp or by >8–10% in voltage clamp. Applying very brief, gentle pulses of pressure to the patch pipette may help lower the access resistance but can also rupture the patch seal.

In some experiments, it is desirable to voltage clamp cells at a membrane potential that is well away from the normal resting potential. For example, to measure currents through NMDA receptors, it is common to clamp the cells at a positive potential, such as +40 mV. Even with Cs-based electrodes, we find that many types of cells do not tolerate being held at positive potentials for more than ~10 seconds at a time. It is usually best to step the cell between a normal holding potential and the positive potential, and apply stimuli during the positive step. Subtraction of traces with and without stimulation may be needed when using such a protocol, as Cs^+ does not completely block all potassium currents, and some time-dependent current may remain.

Data analysis. Patch clamp experiments most commonly generate time-series analog signal recordings (e.g., a one-dimensional array of membrane voltage or current values). These signals are analyzed using a variety of general signal processing techniques as well as less common techniques devised specifically for analysis of neuronal signals. Given a model of the system we are studying (be it a channel, membrane, neuron, or circuit), the objective of any analysis is to measure one or more parameters of the model from signals in the recording. However, the presence of noise and other interfering signals can make this challenging.

Prior to analysis for signals of interest, it is common to digitally filter the recording to remove unwanted noise and offsets. To remove a baseline offset, it is usually sufficient to subtract the mean or median value derived from a quiescent period of the recording. Bessel or Butterworth filters are frequently used to remove both high-frequency noise and low-frequency baseline fluctuations. Any filtering must be applied with caution to avoid altering those aspects of the signal that are to be measured. For example, many filters introduce frequency-dependent phase delays that can affect the measurements of event timing in the signal. Filtering can also generate ringing artifacts in response to rapid changes in the incoming signal or noise spikes. They can also alter the apparent kinetics of rapidly activating currents measured under voltage clamp, such as occurs with voltage-gated sodium and calcium conductances, and some fast synaptic conductances. In general, it is wise to always check that the chosen filtering (or any automated analysis, for that matter) produces the expected results for a set of known inputs, by always checking filtered signals and the subsequent analysis results manually.

Signals of interest in patch clamp recordings can be divided roughly into two categories: evoked events and spontaneous events. Evoked events are somewhat easier to analyze because their timing usually follows a predictable delay with respect to the stimulus (e.g., electric shock, photostimulation). Evoked events include excitatory and inhibitory postsynaptic currents, action potentials, and direct perturbations of the membrane potential or holding current. Such events are analyzed to characterize their shape in some way. For example, sudden shifts in membrane potential may be fit to exponential decay curves to determine their time constant; action potentials are measured for their amplitude, width, after-hyperpolarization depth, rising and falling slopes, and other criteria; and postsynaptic conductances are analyzed for amplitude, latency, rising and falling kinetics, or total charge transfer.

Analysis of spontaneous (or otherwise poorly timed) events requires extra effort because the timing of events must be determined before they can be measured. In some cases, it may be difficult to unambiguously distinguish events from background noise or to separate overlapping events. Numerous techniques for event detection have been developed. Most of these work by filtering the signal such that each event is reduced to a single, sharp spike that can be clearly distinguished from the background noise. The timing of these spikes is then detected by searching for regions of the signal that exceed a predefined threshold. A commonly used and more sophisticated analysis uses a template matching algorithm (Clements and Bekkers 1997). In this approach, a short template with the expected event shape is slid across the trace in time, and the error in the fit (with the baseline and peak amplitude as the adjustable parameters) is returned at each point in time. The regions with the best fits that exceed a statistical criterion are then identified for subsequent analysis. An alternate treatment that can be used in current clamp recordings uses deconvolution to estimate the time course of a current from the voltage traces (Richardson and Silberberg 2008). This method can be useful for isolating and measuring the amplitudes of overlapping events.

Troubleshooting

Slice Viability

Difficulty keeping sliced brain tissue alive is one of the most common problems in slice electrophysiology. The reasons for this are not well understood, and the remedies are often highly specific to each brain region. However, many of the rules for producing viable slices are common across all brain regions:

- Reduce ischemia:
  - Become faster at dissection and slicing.
  - Make sure you are using a well-oxygenated dissection buffer.
  - Try doing the preparation at different temperatures (cooling or warming the tissue during dissection and slicing).
- Reduce excitotoxicity:
  - Use NMDG/sucrose solutions—reducing Na^+ may reduce action potential firing and minimize synaptic release. It may also reduce energy demands.
  - Cut in high-Mg^{2+}/low-Ca^{2+} solutions.
  - Try cooling to reduce activity.
- Check the pH and osmolarity of all solutions:
  - pH should be 7.2–7.4 for all solutions. If needed, supplement buffers with HEPES to help control pH.
  - Reduce bicarbonate to 20 mM in cases where CO_2 content of gas tank may be low.
  - Osmolarity should be about 290 mOsm for electrode solution and 310 for ACSF. Reduce evaporation or add sucrose to make osmolarity appropriate.
- Reduce mechanical trauma:
  - Try different blades for slicing.
  - Cut more slowly if the tissue is not cutting (sticking to blade, rolling, or compressing).
  - Cut more quickly if possible.
  - Check blade angle—the blade should be pointed roughly 10 degrees downward, such that the bottom tapered edge of the blade is horizontal.
  - Try a slice orientation that severs fewer or only smaller processes. For example, some neurons are mostly planar and survive better when the slice is parallel to the main plane of the dendritic tree.
  - Use more care during dissection. Do not touch the region you wish to study, avoid compressing or shocking the tissue. Do not expose the tissue to air any longer than necessary.
  - Cut any cranial nerves before removing the brain from the skull, as the tension from stretching these may damage some areas of the brain.
  - Some slicers impart a small amount of vertical vibration to the blade and may need to be tuned to avoid this.
- Use younger animals; their cells tend to be much more resilient.
- Look in the literature for proven protocols (for your brain region), and talk to people using those regions.
- Be systematic: Start with something that works, and change one variable at a time.
- Look deeper into the slice (may require illumination adjustment).
- Wait at least 1 hour after cutting before starting any recordings.

Because slice preparation and patch clamp have a steep learning curve, we suggest starting with a simple experiment to be sure that you have a handle on how to make everything work together. Start out with younger animals (P10–P14 for rats or mice), and just try to perform current clamp experiments on cells. Once you can regularly get cells with good spike heights and resting potentials, then it is time to advance to your project. Be persistent and expect to spend weeks to months becoming proficient with these techniques.

Electrical Noise

Noise is common in electrophysiology equipment, and the noise both contaminates the recordings and in some cases can mask signals. There are five primary sources of noise. Line noise appears as 50 or 60 Hz, often with harmonics (integer multiples of 50/60 Hz), and can come from several sources, including unshielded power cords, overhead lights, and incorrect grounding practices. Power supply noise is typically also at harmonics of 50 or 60 Hz and may result from power supplies that are not working correctly or are poorly designed or incorrectly grounded. For example, switching power supplies are convenient because they are small and light and do not require heavy transformers or large filter capacitors. However, they can generate wideband noise, anywhere from 15 kHz to many MHz, and this noise may become aliased into the recorded signals. Electrochemical junction noise appears as an unstable, fluctuating noise on a time scale of milliseconds to seconds. It may be caused by salt solution that is spilled, for example near the chamber, that creates a battery between dissimilar metals, or that bridges different ground connections. Digital equipment noise can occur because computers, some modern computer-controlled amplifiers, microprocessor-based devices, and digital and analog cameras have high-frequency clocks or oscillators that are often not properly electrically isolated from ground circuits, and the signals can get into the ground system. Because these are high frequency (sometimes in the MHz range), they can be hard to troubleshoot and may require particular attention. We have found that even top-of-the-line electrophysiology amplifiers can "leak" such signals into the rig and cause problems. Finally, ground loop noise is caused by currents that circulate in the ground system of the rig, through cables and the various grounds associated with the equipment.

To minimize noise, start by stripping the rig down to just the amplifier, the microscope, and the computer. This means disconnecting all cables at both ends and turning off (and unplugging) all other equipment. Remove anything from the vicinity of the microscope that is not being used. Unconnected or dangling cables should be stowed. The microscope body should be grounded to the table, and the table grounded to the Faraday cage, with at least quarter-inch-wide stranded wire strapping of as short a length as practical. All connections should be made using screws and toothed washers, and all surfaces that are to be bonded should have exposed metal at the point of contact (sand off any paint or oxidation). Place a small amount of saline in the recording chamber and a filled pipette on the headstage. Monitor the output of the amplifier with an oscilloscope. We also find that using a spectrum analyzer (or spectrum calculated from digitized data) is extremely helpful in identifying noise sources and eliminating them. The amplifier headstage should only be connected to the electrode, and the high-quality signal ground (usually located on the amplifier headstage) should be connected only to the reference electrode in the bath. In some cases, it is worthwhile either to leave the electrode floating in the air above the recording bath, or to patch a ball of Sylgard in the recording chamber (essentially making a high-impedance seal; under these conditions, small currents are more easily detected). The amplifier itself may have a separate ground connection (on the back), and this can be used as the reference (ground) for the Faraday cage, table, and microscope (and the equipment rack if one is used). With the amplifier filters open (>50-kHz bandwidth), re-examine the noise levels and try to identify and correct noise from any additional sources.

Items that are connected to the microscope can also cause problems and should be addressed next. Any ungrounded conductive objects in proximity to the recording chamber may act as an antenna, which picks up electromagnetic radiation, and couple it capacitively into the recording area. This may include micromanipulators, parts (or all) of the microscope, and the experimenter's body. While some microscopes have a specific ground point that can be used, many parts of the microscope are not bonded electrically to this point and are therefore ungrounded. This often occurs because items are painted or anodized, or because they are separated by a thin layer of grease. Items attached through couplers, such as cameras, should not be considered to be grounded and may need a separate grounding strap. Anything on the stage that is anodized (or is on an anodized stage) will not be well grounded and may need a separate connection.

After eliminating sources of capacitively coupled noise, begin adding the other equipment to the setup one piece at a time, evaluating the noise at each stage, both with the equipment connected and off, and with it turned on. Often this will identify an offending item, which might need to be moved or might need additional attention for grounding.

Ground loop noise can be difficult to eliminate. Ground loops occur when there is more than one path for current to flow between two points in a system. This can occur through shielded cables that connect two pieces of equipment that otherwise share a ground connection with each other. There are also common-mode currents that may flow on a shared signal or ground path that may contribute to interference. There are several treatises on this problem in the professional audio literature (for example, Waldron, Web Resource). There are two points with regard to small rigs. First, maintain as best as possible a "star" ground configuration for all equipment. In a star configuration, there is one central reference point, and all common connections go to that point. While this topology minimizes the chances of creating ground loops between different pieces of equipment, it is not always practical. Second, keep the headstage and its reference input completely separate from the rest of the system grounds (remember also that the interior of the recording chamber and any connecting fluid compartments must be completely electrically insulated from the rest of the system). Third, consider the signal paths associated with connecting cables between equipment items. In some cases, it may be useful to isolate the ground side of the connection in a signal cable, but this is not always recommended. Not all manufacturers follow the same rules for signal grounding in their equipment, and this can cause interesting problems. Sometimes, even short "ground" leads can pick up radiated signals and introduce additional noise. It should never be necessary to disconnect ("lift") the safety electrical ground in a piece of modern equipment if the manufacturer has arranged this correctly (e.g., connected to the equipment case and separate from the signal ground paths). In some cases, where signals >100 kHz are problematic, the use of toroidal cores or ferrite chokes around the connecting cables may be helpful. The size and permeability of the core or choke should be commensurate with the frequency of the signal to be blocked. Sometimes signals from nearby AM or FM radio stations, hospital dispatchers, or even a cell phone or tablet computer in the vicinity of the rig can introduce unwanted energy onto the cables around a rig and add noise. Remember that some of these devices have a wireless connection that operates in the 2- to 5-GHz range, where wavelengths are short, and even a short ground strap or a stray wire can operate as a receptive antenna at these frequencies.

A day spent disassembling and reassembling a rig while monitoring the noise levels can be very helpful to understand how the rig is configured and the various sources of noise in and around the rig. Take a very systematic approach and try to keep the rig as "clean" both physically and electrically as possible; only then will you be rewarded with a low-noise setup whose data traces will make you proud. Regular maintenance, including cleaning the rig and checking the noise level, and maintaining a log of noise measurements under a fixed set of conditions, is also advisable.

As mentioned earlier, another source of noise that sometimes appears is caused by salt spills (even evaporated spills with just salts in a humid environment). If the salt is in a location that can add currents through a ground loop, or create a loop, it can act like an unstable battery. An example is salt bridges between the recording chamber and the metal platform that holds the chamber. Here, the salt creates an unstable resistance possibly with an electrochemical potential between the high-quality ground used in the recording bath (connected to the headstage) and a general ground used for reduction of capacitive noise pickup. For this reason, among others, it is important to clean up all spills immediately. It is also important to take apart and clean any items that may get salt inside them (e.g., microscope, substage condensers, translation stages, manipulators) as soon as possible after a spill. Spills should be carefully cleaned up with water, followed by 70% alcohol, and wiped dry. If spills happen frequently, some items may need to be treated with a thin layer of grease or a rust preventative.

When troubleshooting noise, remember also that the tubing used to bring solutions to and from the preparation contains a conductive solution that can also be capacitively coupled to other noise sources. Sometimes shielding the tubing, or changing its placement, can help. Peristaltic pumps can also introduce noise through the fluid delivery system, and should be avoided when possible. However, peristaltic pumps are sometimes needed when using expensive or limited chemicals in a recirculating bath.

Vibration

Vibration isolation tables are designed to dampen vibrations that commonly occur in buildings, usually in a low-frequency range that depends on the size of the table. If the electrode is vibrating under the microscope, then there may be a mechanical connection that is essentially short-circuiting the isolation table. Anything that goes on or off the table can contribute to this. Whenever possible, use cables that are flexible to bring signals to and from devices on the table, and clamp (or tape) the cables to the edge of the table where they leave. Allow the cables to hang (do not make them tight) so that vibration from other nonisolated instruments and racks is less well coupled to the table. Devices with fans, such as some high-performance CCD cameras, can also contribute to vibration, especially if they are mounted at the top of the microscope. In extreme cases it may be necessary to replace the fans, manually balance them, or find a way to mechanically uncouple the camera from the microscope.

Frequently Asked Questions

Q: I don't see any living cells in my slice.

A: This is one of the most common problems in slice electrophysiology (see "Troubleshooting: Slice Viability"). The first thing to remember is that the ideal dissection, slicing, and incubation procedures vary considerably between brain regions, so find out what has already worked for other researchers slicing the same region. If you believe you are doing everything correctly, try cutting slices from younger animals, which are typically much easier to work with. Think critically about your procedure and remember the major causes of cell death in brain slices: pH, ischemic damage, excitotoxic damage, and mechanical damage. Consider also that a dead slice should have many cells that appear to be dead, whereas a slice with no cells may simply indicate an illumination or imaging problem. Finally: Be persistent.

Q: I am not able to form a gigaohm seal.

A: The most common cause is that the tip of the pipette is fouled. This can have several causes: (1) There were fingerprints or dust on the center of the pipette glass before pulling, (2) Pipettes are too old (more than a day) or left uncovered too long, (3) The pipette contacted crystallized salt on the surface of the recording chamber water, (4) The pipette tip contacted brain tissue before it reached the cell due to insufficient pressure inside the pipette, or (5) The bathing solution contains serum or bovine serum albumin. If you have to use a solution that contains these, make the seal first in a solution that does not contain proteins, and then switch solutions.

Two other reasons: (1) A dimple was not visible on the cell before releasing pressure (this often means that something else was compressed between the pipette and the target cell) or (2) The cell is dead.

Q: I get a gigaohm seal very quickly, but the cell seems to be gone immediately after breaking in (indicated by very low input resistance or very high resting membrane potential).

A: The cell was probably dead before you patched it. Cells that look similar to this one are also likely to be dead; try changing your cell selection criteria.

Q: I can patch a cell, it looks healthy, but I lose it 10 minutes later.

A: This is often caused by a drifting pipette. Check to see that it has not moved more than ~10 μm from its location at the time of patching. If the pipette has moved, see the following question.

Another possibility is that vibrations transmitted to the pipette tip caused it to detach from the cell. This can be caused by vibrating equipment (unbalanced camera fans are a common culprit), poor isolation from ground vibrations, or the experimenter touching the setup. It is recommended to monitor the cell's health frequently (or use an audible indicator of electrode potential) to increase the probability that you will discover the cause of a lost cell.

If this happens consistently, it is possible that your internal solution is poisoning the cell. If the cell becomes swollen or shriveled after patching, it is possible that the osmolarity of your internal solution is too low or too high. Try using a different aliquot of internal solution, or borrowing an aliquot of a different batch of internal solution from another experimenter to see if that solves the problem.

Q: My pipette tip is drifting! What do I do?

A: This is most commonly caused by overtightening or undertightening the electrode holder cap, which compresses and strains the O-ring holding the electrode. The O-ring relaxes slowly over time, causing the electrode to drift. Applying a small amount of grease to the O-ring can help release this strain before it becomes a problem. Also be sure that nothing is touching the pipette, such as the edge of the recording chamber or the objective, and that the pressure tube and headstage cable are properly secured to prevent transmission of any strain to the electrode holder. Another source of drift can be temperature changes in the vicinity of the headstage, or a malfunctioning manipulator.

Q: My patch pipettes keep clogging.

A: Clogged pipettes are a common but easily solvable problem. These are most commonly caused by either particulates suspended in the electrode solution or a dirty pipette filler. Electrode solutions should either be centrifuged at the beginning of the day or filtered immediately before filling the pipette, or both. We recommend against using most commercial pipette fillers, as they are difficult to clean. Instead make new fillers daily from plastic pipettes.

Q: My ACSF solution looks cloudy or has a precipitate.

A: This is often a sign that the pH of the solution is not in the right range (7.2–7.4). Check the pH. If the solution is being gassed with carbogen, check to be sure that the tank really contains 95% O_2 and 5% CO_2. One of the authors had an experience where the supplied tank did not have any CO_2 in it, and this rapidly led to the death of the slices.

Q: My series resistance (or bridge balance) starts out OK but increases over the course of the experiment.

A: Series resistance should be kept to a minimum (<15–20 MΩ) when possible. Increased series resistance is usually a sign that the tip of the pipette is clogging or the membrane is resealing. Sometimes a little pressure or suction will reopen the tip and allow access. It may also be a sign that the pipette is drifting away from the cell. In this case, visually check the pipette position and check the "My pipette tip is drifting!" FAQ. Increasing the diameter of your pipette tips can help avoid this problem.

Q: I can't seem to compensate the amplifier in voltage clamp.

A: First, be sure that you understand the compensation procedure in the manufacturer's manual, and practice the procedure on a model cell. Second, make sure that the electrodes are properly coated to reduce capacitance and that the electrode series resistance is low (<5–10 MΩ). Make sure that the holding potential and the voltage step are in a linear range for the cell you are working with. The activation and deactivation of voltage-dependent channels can happen on the same time scale as the transients that you are trying to eliminate, making it difficult to properly adjust the compensation. Finally, remember that neurons with extensive dendrites do not appear to the clamp amplifier as a simple resistor-capacitor circuit with a single time constant but have a large number of time constants. However, the amplifiers only are designed to provide proper compensation for a cell with a single time constant. In voltage clamp, you should be attempting to compensate the fastest time constant and will not be able to correct for the slow components.

Q: There is a high-frequency intermittent spiky noise on my recordings.

A: Turn off your cell phone and pager, or move them away from the rig. Make sure that the aspiration of solution from the chamber is not causing a charge separation (sometimes using a fine silver wire for the first 10–20 cm in the aspiration tube will help keep this from happening).

Specific reagents are also necessary (Table 1.3). Salts should be purchased from a reliable vendor and should be at least ACS grade or better. In particular, pay attention to the level of impurities in the salts. Storage should be according to the vendor's recommendations.

TABLE 1.3 Necessary Equipment

Equipment	Potential Sources
SLICE AND PATCH SPECIFIC EQUIPMENT	
Vibration isolation table	Newport, Kinetic Systems, Technical Manufacturing Corp.
Pipette puller	Sutter Instruments, Narishige, Warner
Upright fixed-stage microscope	Zeiss, Nikon, Olympus
Hot air gun	
Binocular (stereo) dissecting microscope	Zeiss, Nikon, Olympus
Tissue slicer and light source	Leica, Pella, Camden, FHC
Recording amplifier	Axon/Molecular Devices, Heka, Dagan, NPI, Alembic Instruments
Micromanipulator	Sutter, Burleigh, Luigs & Neumann
CCD camera	
Oscilloscope (for troubleshooting)	
Slice chamber	Warner Instruments, Scientific Systems Design, Cell MicroControls
Stage temperature controller	Warner Instruments
Translation stage	Sutter, Burleigh, custom designs
Faraday cage (optional but recommended)	
Computer	
Data acquisition system or cards	Molecular Devices Digidata for use with pClamp, National Instruments, Data Translation, Cambridge Electronic Design
STANDARD LABORATORY EQUIPMENT	
Balance	
pH meter	
Osmometer	
Pipettors	
Refrigerators, freezers	
Centrifuge	Eppendorf
Water bath	
General glassware (flasks, beakers, graduated cylinders)	
Sintered glass gas dispersion tubes	Corning
Tygon and Teflon tubing	
Silastic tubing	
Polyethylene tubing (for pipette suction line)	
Teflon valves	Cole Parmer
Pipette glass	Garner glass co. KG33 or N51, Sutter Instruments 1.2 mm pre-polished, WPI, Dagan, AM systems
Holding chamber	

(continued)

TABLE 1.3 *(Continued)*

Equipment	Potential Sources
SOFTWARE	
Data acquisition software	pClamp, Cambridge Electronic Design, or custom such as Acq4 (Campagnola et al. 2014)
Data analysis software	Igor Pro (Wavemetrics Inc. Oswego, OR) with TaroTools (https://sites.google.com/site/tarotoolsregister/) or Neuromatic (http://www.neuromatic.thinkrandom.com), pClamp (Molecular Devices), AxoGraph (http://www.axograph.com), MATLAB (The Mathworks, Natick, MA)

We have indicated our preferred storage below (D = desiccator, R = refrigerated at 4°C, F = freezer at –20°C, F80 = freezer at –80°C). Desiccation may not be a problem if you live in a dry area such as the Southwestern desert, but in the American South, it can be important.

Reagents: NaCl (D), KCl (D), KH_2PO_4 (D), $MgSO_4$ (D), $CaCl_2$ (D), Glucose (D), Sucrose (D), N-Methyl-D-Glucamine (D), Ascorbic Acid (R), Myoinositol (R), Sodium pyruvate (R), HEPES (D), EGTA (D), K-gluconate (D), Mg-ATP (F80), Na-GTP (F80), phosphocreatine (F80), Alexa-fluor 488 or similar dye, hydrazide salt (F), Lucifer Yellow (K^+ salt) (F).

Acknowledgments

This work was supported by NIH Grants DC004551 and DC009809 to PBM.

References

The Axon Guide: A Guide to Electrophysiology & Biophysics Laboratory Techniques (3rd ed.) (2008). Sunnyvale, CA: Molecular Devices/MDS Analytical Technologies.

Blanton, M. G., Lo Turco, J. J., & Kriegstein A. (1989). Whole cell recording from neurons in slices of reptilian and mammalian cerebral cortex. *J Neurosci Meth* 30: 203–210.

Brown, K. T., & Flaming, D. G. (1995). *Advanced Micropipette Techniques for Cell Physiology*. IBRO Handbook Series: Methods in the Neurosciences (V9). New York, NY: John Wiley and Sons..

Campagnola, L., Kratz, M. B., & Manis, P. B. (2014). ACQ4: an open-source software platform for data acquisition and analysis in neurophysiology research. *Frontiers in Neuroinformatics* 8: 3.

Clements, J. D., & Bekkers, J. M. (1997). Detection of spontaneous synaptic events with an optimally scaled template. *Biophys J* 73: 220–229.

Hamill, O. P., Marty, A., Neher, E., Sakmann, B., & Sigworth, F. J. (1981). Improved patch-clamp techniques for high-resolution current recording from cells and cell-free membrane patches. *Pflugers Arch* 391: 85–100.

Lorteije, J. A., Rusu, S. I., Kushmerick, C., & Borst, J. G. (2009). Reliability and precision of the mouse calyx of Held synapse. *J Neurosci* 29(44): 13770–13784.

Magistretti, J., Mantegazza, M., de Curtis, M., & Wanke, E. (1998). Modalities of distortion of physiological voltage signals by patch-clamp amplifiers: a modeling study. *Biophys J* 74: 831–842.

Neher, E., Sakmann, B., & Steinbach, J. H. (1978). The extracellular patch clamp: a method for resolving currents through individual open channels in biological membranes. *Pflugers Arch* 375: 219–228.

Richardson, M. J., & Silberberg, G. (2008). Measurement and analysis of postsynaptic potentials using a novel voltage-deconvolution method. *J Neurophysiol* 99: 1020–1031.

Rothman, J. S., & Manis, P. B. (2003). The roles potassium currents play in regulating the electrical activity of ventral cochlear nucleus neurons. *J Neurophysiol* 89: 3097–3113.

Shao, X. M., & Feldman, J. L. (2007). Micro-agar salt bridge in patch-clamp electrode holder stabilizes electrode potentials. *J Neurosci Methods* 159: 108–115.

Snyder, K. V., Kriegstein, A. M., & Sachs, F. (1999). A convenient electrode holder for glass pipettes to stabilize electrode potentials. *Pflugers Arch* 438: 405–411.

Smith, T. G., Lecar, H., Redman, S. J., & Gage, P. W. (Eds.) (1985). *Voltage and Patch Clamping with Microelectrodes*. Bethesda, MD: American Physiological Society.

Spruston, N., Jaffe, D. B., Williams, S. H., & Johnston, D. (1993) Voltage- and space-clamp errors associated with the measurement of electrotonically remote synaptic events. *J Neurophysiol* 70: 781–802.

Tanaka, Y., Tanaka, Y., Furuta, T., Yanagawa, Y., & Kaneko, T. (2008). The effects of cutting solutions on the viability of GABAergic interneurons in cerebral cortical slices of adult mice. *J Neurosci Methods* 171: 118–125.

Waldron, T. *A Practical Interference-Free Audio System (parts 1 and 2)*. Web resource: http://www.nutwooduk.co.uk/archive/Old_Archive/020918.htm

Williams, S. R., & Mitchell, S. J. (2008) Direct measurement of somatic voltage clamp errors in central neurons. *Nat Neurosci* 11: 790–798.

Yamamoto, C., & McIlwain, H. (1966) Potentials evoked in vitro in preparations from the mammalian brain. *Nature* 210: 1055–1056.

Patch Clamp Recording *in Vivo*

David Ferster

Introduction

The study of intracellular membrane potentials has a surprisingly long history. Sharp micropipettes—originally developed for injecting material into cells—were combined with high-impedance amplifiers to record membrane potentials from plants and unicellular organisms as early as the 1920s (for review, see Stuart and Brownstone 2011). Intracellular recordings of membrane potential in mammalian cells were first obtained from muscle fibers (Ling and Gerard 1949), and recordings from neurons soon followed with the concurrent work of Eccles (1952) and of Woodbury and Patton (1952) in motoneurons of the cat spinal cord. Until the advent of patch recording, sharp electrode recording remained the only method to give direct access to the membrane potential of single mammalian neurons *in vivo*. And while sharp electrode recording made possible a number of profound discoveries in central nervous system physiology, it remains a daunting technique, even when applied to neurons as large as motoneurons.

The main difficulty with sharp electrode recording *in vivo* relates to stability. When the electrode penetrates a cell, the membrane meets the surface of the electrode at more or less a right angle. There is little opportunity for the membrane to seal to the surface of the electrode, and any leaks will lead to depolarization of the neuron and will degrade the quality of recording. The recording is therefore highly sensitive to disruption by the inevitable movements that occur in the brain of a living animal, from the heartbeat and respiration and, if the animal is not paralyzed, from muscular contractions.

Patch clamp recording goes a long way toward solving this problem. In a patch recording, the membrane is drawn up into the electrode for a fraction of a micron, making the well-known omega figure that leads to a gigaohm seal. The membrane surface lies parallel to the inner wall of the glass and can form a physically strong and electrically tight bond with the glass that is not easily disrupted by small movements and is resistant

(a)

(b)

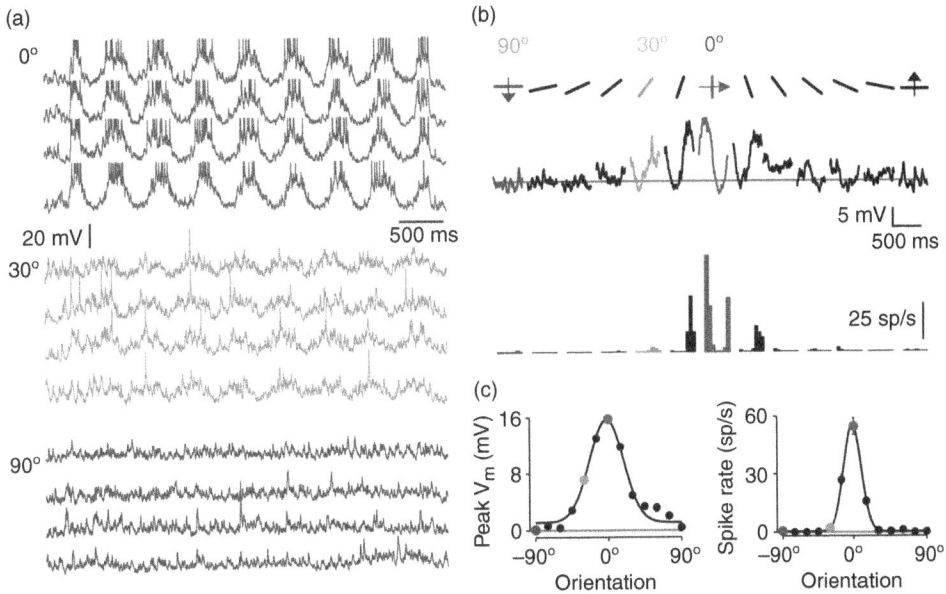

FIGURE 2.1 **Intracellular records obtained from a patch recording from a simple cell in primary visual cortex of the anesthetized cat.** A: Membrane potential fluctuations evoked by a moving sinusoidal grating presented in the receptive field of the cell at three different orientations (relative to the cell's preferred orientation). Depolarizations at 0 degrees occurred as each bar of the grating crossed the receptive field. B: For each of the 13 orientations of the grating, the average response to each bar of the grating was calculated for V_m (above, with spike removed) and spike rate (below). C: The peak-to-peak amplitude of the V_m responses and peak spike rate are plotted against stimulus orientation.

to electrical leaks. These features make patch recording ideal for use *in vivo*, and it was soon adopted for intracellular recording from the visual cortex, initially with incomplete gigaseals (Pei et al. 1991) and subsequently with full gigaseals (Jagadeesh et al. 1992).

Patch recording has now become a reliable and accessible tool for obtaining high-quality intracellular recording from neurons—regardless of size—in parts of the mammalian brain as diverse as the visual cortex (Figs. 2.1 and 2.2), auditory cortex (Wehr and Zador 2003), somatosensory cortex (Bruno and Sakmann 2006, Ganmor et al. 2010), thalamus (Wang et al. 2011), basal forebrain (Metherate and Ashe 1993), olfactory bulb (Margrie et al. 2001), spinal cord (Graham et al. 2004), olivary nuclei (Bazzigaluppi et al. 2012), and cerebellum (Rancz et al. 2007). The low access resistance of the whole-cell configuration makes it much easier than in sharp recordings to perfuse the recorded cell with different ion solutions, pharmacological agents (Jia et al. 2010, Nelson et al. 1994), dyes for two-photon calcium imaging (Jia et al. 2010), and viral vectors for retrograde synaptic tracing (Rancz et al. 2011). Patch recording makes it possible in some cases to apply single-electrode voltage clamp to measure membrane current and conductance (Tan et al. 2011, Zhang et al. 2003). Given its ability to withstand considerable movement of the brain, patch recording is being adapted for use in recording from awake animals (Fukunaga et al. 2012, Okun et al. 2010, Tan et al. 2014).

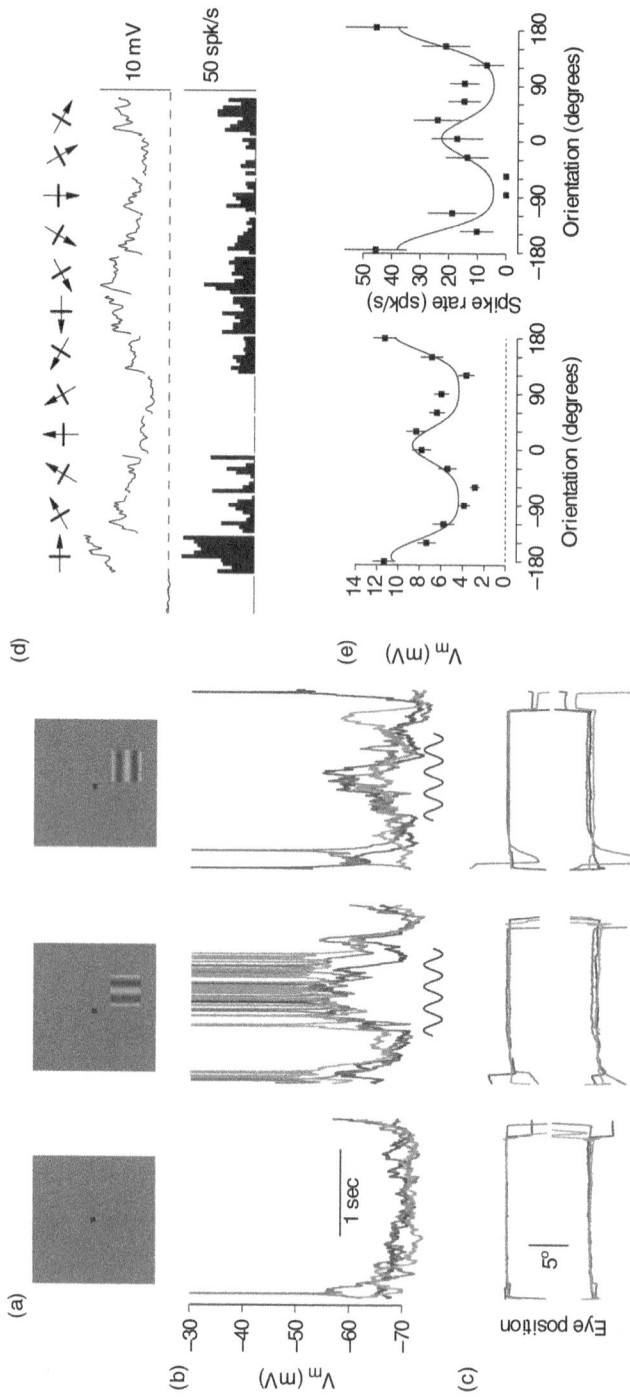

FIGURE 2.2 Intracellular records obtained from a patch recording from a complex cell in the primary visual cortex of an awake monkey. A: Fixation point and visual stimuli at two different orientations. **B:** Superimposed responses to a blank screen and to the preferred and orthogonal orientation. **C:** Vertical (above) and horizontal (below) eye position during each trial. **D:** Averaged membrane potential and spike response to stimuli at 12 different orientations. **E:** Tuning curves derived from the averaged responses in D. (From Nicholas Priebe.)

The key developments that made *in vivo* patch recording possible were (1) the original patch recording method for acutely isolated or cultured cells (Hamill et al. 1981) and (2) the adaptation of the patch recording technique for blind patching in brain slices (Blanton et al. 1989). In the original patch recording methods for isolated or cultured cells, Hamill et al. guided the electrodes to the surface of cell visually. Before the routine adoption of differential interference contrast microscopy, however, it was difficult to pick out individual cells in brain slices and target patch electrodes toward them. Blanton et al. (1989) therefore relied on there being a high enough density of cells in the slice that the electrode would randomly encounter the soma of a cell after advancing the electrode a few hundred microns. To detect when the electrode was approaching a cell, they recorded the increase of the electrode resistance that occurs as the cell membrane starts to block the electrode tip.

A second difference between patching isolated cells and patching cells in slices is that in slices, the neuropil and vessels surrounding the cells tend to block the electrode as it is advanced through the tissue. The same problem can occur *in vivo*. In addition, in the intact brain, pushing the electrode through the pia mater can also clog the electrode, although this problem is more severe in cat and other large animals than in small animals such as rodents. To mitigate clogging, slight pressure is maintained on the electrode during the approach to the cell. The pressure ejects a small amount of solution, helping to keep the electrode tip clear. In addition, as the electrode approaches the membrane of a cell about to be patched, the flow of liquid likely cleans off the membrane surface somewhat to facilitate a tight seal (Coleman and Miller 1989).

Two additional steps are needed to adapt the blind patch technique for use *in vivo*. First, the electrodes are given a longer (more gradual) taper. Normally, the electrode is designed to minimize the resistance of the body of the electrode away from the tip. This is more important for patch electrodes than for sharp electrodes since the electrode solution is so much more dilute (300 mM compared to 3M or greater) and therefore more resistive. Patch electrodes are therefore usually given very steep (short) tapers so that the liquid path from the silver wire to the tip is as wide as possible for as long as possible. Extreme tapers have no disadvantages for use with dissociated cells and can also be used effectively in slices since the typical distance between the surface of the slice and the recorded cell is 200 μm or less, and the electrode enters the tissue through the rather forgiving cut surface of the slice.

In vivo, however, the electrode enters through the pia mater and cells can be recorded at a depth of 1,000 μm or greater. For both of these reasons, traditional patch electrodes with very steep tapers cause considerable dimpling or compression of the surface of the brain. This problem becomes even more severe if the electrode is placed at a low angle with respect to the surface of the brain in order to fit underneath a microscope objective for two-photon microscopy. To reduce dimpling and catching on the pia as much as possible, therefore, the electrodes are given a much longer taper. For work in the cat cerebral cortex, we use electrodes that taper for 7–8 mm.

The second modification of the blind patch technique for *in vivo* recording is to use slightly smaller electrode tips than those used in slices or dissociated cells. Smaller tips seem to clog less during the traverse through overlying tissue to the recorded cell. Clogging of the electrode tip by neuropil becomes more of an issue when recording from deeper and deeper cells and also seems to be more severe in cats than in rodents. While smaller tips may reduce clogging, their use must be weighed against the resulting increase in the access (series) resistance of the electrode. With 1.2-mm glass, more gradual tapers, and smaller tips, our electrodes usually fall in the range of 5–8 MΩ (before entering the brain). Lower resistances can be used when recording currents is the goal, but potentially at a cost of a lower success rate in obtaining recordings.

Detailed Methods

Electrodes

Electrodes are pulled from thin-walled borosilicate glass pipettes with an outer diameter of 1.2–1.5 mm and an internal fiber (omega dot glass). We use a two-stage pull on a puller from Sutter Instruments, equipped with a 3-mm box filament. The final electrodes have tapers of 7–8 mm in length. The last 0.5 mm of the taper appears very linear when examined at low power under a compound microscope. The tip viewed under high power is about 1 μm outer diameter. When filled with solution, the electrodes have resistances in the range of 7 MΩ.

Note that using thin-walled glass keeps the electrode resistance down but has the disadvantage of increasing the capacitance per unit length of the electrodes. When recording *in vivo*, the electrode is driven relatively deep into the brain, and in addition agar is often placed over the craniotomy. With several millimeters of the electrode in contact with brain or agar, the high capacitance per unit length is compounded to give a high overall capacitance of the electrode. The time constants of the electrodes increase accordingly. Time constants can be critical when trying to perform voltage clamp experiments, since good clamp for short-duration events depends on having a low electrode time constant. A long time constant can also distort the recorded action potential waveforms, making them broader and shorter. As noted below, good results can also be achieved with thick-walled glass, which for these reasons might be preferable.

Intracellular Solution

The electrode solution is standard to patch recording and introduced into the electrode through the back. For high-resistance, long-taper electrodes, we immerse the back end of the electrode in a small container of solution for a few minutes to allow the solution to wick into the tip along the internal fiber, and then fill the rest of the electrode using a 1-mm syringe and fine needle (30 gauge) or drawn plastic pipette tip. For the solution, we use (in mM) 135 K-gluconate, 4 KCl, 10 HEPES, 0.5 EGTA, 10 Na_2-phosphocreatine, 4 Mg-ATP, and 0.4 Na_2-GTP, adjusted to pH 7.3 and 285–290 mOsm. For recording purely synaptic responses in voltage or current clamp mode, active Na^+ and K^+ currents can be blocked by adding QX-314 to the electrode (5 mM) and substituting Cs^+-methanesulfonate for K^+-gluconate. The Cs^+ substitution, however, can make obtaining gigaseals more difficult. Solutions containing nucleotides should be kept frozen until use or prepared fresh the day of use.

Stability

One issue that arises in *in vivo* recording that is not relevant to recordings in brain slices is the stability of the brain. In mice, it appears that no special measures are required to stabilize the brain other than to place warm agar around the electrode. This also helps to keep the surface of the brain moist during recordings. In cats, however, movement of the brain from the heartbeat, and in particular from the large-amplitude, low-frequency oscillations associated with respiration, can disrupt the recordings. For this reason, we routinely perform a pneumothoracotomy prior to recording, a procedure that has long been common in experiments using sharp electrodes *in vivo*. A small tube inserted into the incision covered at the open end with a small balloon helps keep the intrathoracic space moist. Preventing the intrathoracic pressure from oscillating with respiration in this way greatly reduces movement of the brain. Although we have never done so, introducing an opening in the dura near the brainstem to reduce the pressure of the cerebrospinal fluid is also possible.

For awake recordings, neither a pneumothoracotomy nor draining cerebrospinal fluid is an option. Instead, it is possible to place a pressure foot (a small, rigidly held metal plate covering the exposed portion of the cortex) against the surface of the brain and introduce the electrodes through a small hole in the plate. Slight pressure on the surface of the brain seems not to affect the health of the underlying tissue but provides considerable stability, even in a behaving animal (Nicholas Priebe, personal communication).

Equipment

The equipment for *in vivo* patch recording is nearly identical to that required for patch recording in slices, including a patch clamp amplifier, data acquisition interface (with analog-to-digital and digital-to-analog converters, timers, and digital input and output), computer, and data acquisition software. An oscilloscope is often useful to view signals in real time, although it is not required. Depending on the species of animal used, an appropriate stereotaxic head-holder is needed to keep the head fixed. Additional equipment for animal maintenance, anesthesia, sensory stimulation, or behavior control will vary depending on the experiment. Electrodes are positioned and advanced using motorized manipulators standard to *in vitro* patch recording.

Recording

Electrodes are placed with the tip just above the surface of the brain and the craniotomy is flooded with a layer of warm agar (3% in normal saline). At the beginning of the electrode track, before the electrode has entered the brain, we begin passing 100-pA, 100-ms square current pulses through the electrode. The amplifier's bridge is balanced to compensate for series resistance and the electrode resistance is noted. Positive pressure (200–300 mbar) is applied to the electrode with a small syringe connected by silicon tubing to the rear of the electrode, and the electrode is then advanced into the brain and on to the region or layer of interest. Once in the region of interest, the pressure is reduced to 20–40 mbar, and the electrode is advanced a few microns at a time while the response to the current pulse is monitored on the oscilloscope or computer. When the electrode tip nears a cell, the resistance will rise and fall in a characteristic manner, rising slightly as the electrode is moved forward, then falling partway back to its previous value when the motion stops. With several successive moves, the resistance will ratchet up as much as 5 or 10 MΩ.

When these small increases in resistance start to saturate, the pressure is released from the electrode. In the best cases, a gigaseal develops almost immediately. If not (which is often the case), small pulses of negative pressure (<50 mbar) can be applied to the electrode by mouth while monitoring the current pulse. When a successful gigaseal is obtained, the response to the current pulse will grow to saturate the amplifier and its rise time will slow as the enormous increase in resistance increases the electrode time constant.

If suction on the back of the electrode still does not result in a gigaseal, it is possible to continue on with the same electrode and try again on a different cell. Success is likely only if brief pressure on the electrode first clears the tip and reduces the resistance back to its original value. Even then, the likelihood of obtaining a good recording is reduced, and we most often simply replace the electrode after an initial attempt and start again.

Once a gigaseal is obtained, small pulses of negative pressure (100 mbar) are applied by mouth to break the seal and obtain a whole-cell recording, which is indicated by a sudden drop in resistance to the 50–200 MΩ range, an increase in noise (from background synaptic potentials), and the appearance of a resting potential and, possibly, of action potentials. Capacitance compensation is adjusted at this point.

Note that the whole patching procedure may also be performed in voltage clamp mode (as is often done in isolated cells or brain slices), where the current pulse is replaced with a voltage pulse (1–10 mV) and the approach to a cell or occurrence of a seal is indicated by a decrease in the current response to <1–10 pA.

Once the whole-cell configuration is obtained, recording may begin with one caution. During the approach to the cell, the positive pressure on the electrode ejects some volume of electrode solution. Since the solution is high in K$^+$ it can depolarize and silence action potentials in nearby cells and in the one being recorded. The lower the electrode resistance, and the higher the pressure applied during approach, the more the cells can be affected. Depending on the experiment, therefore, it may be desirable to wait some time after a recording is obtained to allow the K$^+$-rich solution to perfuse away and for the extracellular fluid around the cell to return to normal. This may be particularly important if the seal is maintained in cell-attached mode for some time in order to record spikes (before the internal perfusion by the electrode solution perturbs spike generation).

Whole-cell recordings obtained in this manner will often last for an hour or more with little change in the access resistance, resting potential, or apparent health of the cell.

Conductance Measurements

Prior to the advent of patch recording, one of the few ways to apply voltage clamp *in vivo* was to penetrate a cell with two sharp electrodes glued together (Araki and Terzuolo 1962), one for passing current and one for recording potential. This extraordinarily difficult method was only possible in very largest of neurons, particularly the motoneurons of the cat. The advent of patch recording and the single-electrode voltage clamp made voltage clamp in mammalian cells in slices much more accessible. And, with a few adaptations, it is possible to measure changes in membrane conductance *in vivo* as well. As with slice recording, the series resistance of the electrode must be low enough (50 MΩ or less), and care should be taken to compensate the series resistance to reduce clamp errors. Synaptic currents *in vivo* can reach several hundred nanoamps or more, and if the uncompensated series resistance is >20 MΩ, clamp errors of 10 mV or more can result. Series resistances in this range will also create electrode time constants that severely attenuate rapid transients in conductance. Voltage clamp *in vivo*, however, can be effective in measuring relatively slow changes in conductance, such as the responses to low-frequency visual stimuli.

When the electrode series resistance is high, an alternative to voltage clamp is to record in current clamp mode and deliver steady currents to the electrode. This method was first applied by Coombs et al. (1955) using sharp electrodes in motoneurons of the spinal cord, and it is equally effective with patch electrodes. With depolarizing current, the membrane potential can be brought near the reversal potential for excitatory conductances, in which case most the changes in membrane potential evoked by a stimulus can be attributed to inhibition. Conversely, hyperpolarizing current will bring the membrane potential close to the reversal potential for inhibition, such that the remaining membrane potential changes can be attributed primarily to excitation.

Note that this method as described is rather qualitative. Moving the membrane potential near to the reversal potential for inhibition gives records that are enriched for EPSPs, but not exclusively so. A more quantitative version of the current clamp method for revealing excitatory and inhibitory conductances requires injecting steady currents of several levels and delivering the stimuli repeatedly at each level (Priebe and Ferster 2005). From the resulting dataset, a graph of injected current versus voltage is constructed at each time point during the stimulus. The slope of this I-V curve at each point in time gives the conductance of the membrane, changes in which indicate the induction of synaptic potentials. These changes in conductance can be further broken down into underlying changes in excitatory and inhibitory conductance by assuming reversal potentials for the excitatory and inhibitory components of the currents.

Note that in voltage clamp, active currents are presumably not recruited by any stimuli if the membrane potential is truly clamped (i.e., if series resistance is sufficiently compensated). In current clamp, however, the membrane potential is allowed to vary even if it is polarized from rest with steady currents. It cannot be assumed, then, that active currents do not contribute to the observed changes in potential. For this reason, it is critical to block active currents pharmacologically, by introducing QX-314 and Cs^+ into the recording electrode. The extent to which active currents contribute to the recorded fluctuations in membrane potential can also be determined by measuring how linear the I-V curves are. An example of visually evoked conductance changes calculated in this manner is shown in Figure 2.3, from Priebe and Ferster (2005).

A second limitation of the current clamp method is that the series resistance must be compensated carefully while passing current so that the real membrane potential can be determined accurately. This can be achieved offline. Third, both the current clamp and voltage clamp methods are subject to space clamp errors. That is, the membrane potential at the soma is not necessarily the same as the potential in the dendrites, depending on the length constant of the dendrites relative to their physical length. Given these limitations, if applied to high-quality recordings, current measurements *in vivo* can give useful information on the synaptic inputs received by neurons in intact neural circuits.

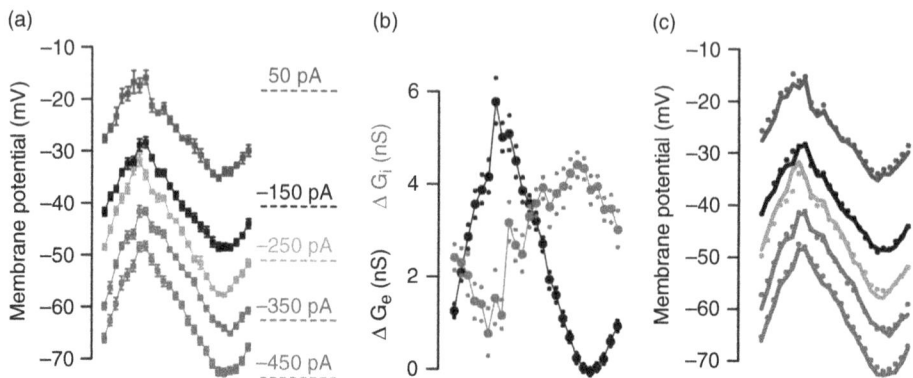

FIGURE 2.3 Current clamp dissection of excitatory and inhibitory currents evoked by a visual stimulus. A: Averaged responses to one cycle of a drifting grating recorded with five different levels of current. B: Time course of the excitatory and inhibitory conductances underlying the voltage responses, as derived from the voltage traces from the membrane equation: $C_m \, dV/dt = g_e(V_m - V_e) + g_i(V_m - V_i) + g_{rest}(V_m - V_{rest}) + I_{inj}/(g_e + g_i + g_{rest})$ where V_e and V_i are the reversal potentials for excitation and inhibition. For the relatively slow responses to the visual stimulus, dV/dt is effectively 0. The five levels of injected current give five equations for finding the best-fit values of g_e and g_i at each point in time. C: The membrane potential records (*solid curves*), together with a reconstruction of the membrane potential records from the conductance estimates (*points*). That the two extracted parameters, g_e and g_i, give a close fit to all five records indicates that the records are dominated by the synaptic conductances and are not distorted by the presence of active conductances.

References

Araki, T., & Terzuolo, C. A. (1962). Membrane currents in spinal motoneurons associated with the action potential and synaptic activity. *J Neurophysiol* 25: 772–789.

Bazzigaluppi, P., Ruigrok, T., Saisan, P., De Zeeuw, C. I., & de Jeu, M. (2012). Properties of the nucleo-olivary pathway: an in vivo whole-cell patch clamp study. *PLoS One* 7: e46360.

Blanton, M. G., Lo Turco, J. J., & Kriegstein, A. R. (1989). Whole cell recording from neurons in slices of reptilian and mammalian cerebral cortex. *J Neurosci Methods* 30: 203–210.

Bruno, R. M., & Sakmann, B. (2006). Cortex is driven by weak but synchronously active thalamocortical synapses. *Science* 312: 1622–1627.

Coleman, P. A., & Miller, R. F. (1989). Measurement of passive membrane parameters with whole-cell recording from neurons in the intact amphibian retina. *J Neurophysiol* 61: 218–230.

Coombs, J. S., Eccles, J. C., & Fatt, P. (1955). The specific ionic conductances and the ionic movements across the motoneuronal membrane that produce the inhibitory post-synaptic potential. *J Physiol (Lond)* 130: 326–373.

Eccles, J. C. (1952). The electrophysiological properties of the motoneurone. *Cold Spring Harb Symp Quant Biol* 17: 175–183.

Fukunaga, I., Berning, M., Kollo, M., Schmaltz, A., & Schaefer, A. T. (2012). Two distinct channels of olfactory bulb output. *Neuron* 75: 320–329.

Ganmor, E., Katz, Y., & Lampl, I. (2010). Intensity-dependent adaptation of cortical and thalamic neurons is controlled by brainstem circuits of the sensory pathway. *Neuron* 66: 273–286.

Graham, B. A., Brichta, A. M., & Callister, R. J. (2004). In vivo responses of mouse superficial dorsal horn neurones to both current injection and peripheral cutaneous stimulation. *J Physiol* 561: 749–763.

Hamill, O. P., Marty, A., Neher, E., Sakmann, B., & Sigworth, F. J. (1981). Improved patch-clamp techniques for high-resolution current recording from cells and cell-free membrane patches. *Pflugers Arch* 391: 85–100.

Jagadeesh, B., Gray, C. M., & Ferster, D. (1992). Visually evoked oscillations of membrane potential in cells of cat visual cortex. *Science* 257: 552–554.

Jia, H., Rochefort, N. L., Chen, X., & Konnerth, A. (2010). Dendritic organization of sensory input to cortical neurons in vivo. *Nature* 464: 1307–1312.

Ling, G., & Gerard, R. W. (1949). The membrane potential and metabolism of muscle fibers. *J Cell Physiology* 34: 413–438.

Margrie, T. W., Sakmann, B., & Urban, N. N. (2001). Action potential propagation in mitral cell lateral dendrites is decremental and controls recurrent and lateral inhibition in the mammalian olfactory bulb. *Proc Natl Acad Sci USA* 98: 319–324.

Metherate, R., & Ashe, J. H. (1993). Ionic flux contributions to neocortical slow waves and nucleus basalis-mediated activation: whole-cell recordings in vivo. *J Neurosci* 13: 5312–5323.

Nelson, S., Toth, L., Sheth, B., & Sur, M. (1994). Orientation selectivity of cortical neurons during intracellular blockade of inhibition. *Science* 265: 774–777.

Okun, M., Naim, A., & Lampl, I. (2010). The subthreshold relation between cortical local field potential and neuronal firing unveiled by intracellular recordings in awake rats. *J Neurosci* 30: 4440–4448.

Pei, X., Volgushev, M., Vidyasagar, T. R., & Creutzfeldt, O. D. (1991). Whole cell recording and conductance measurements in cat visual cortex in-vivo. *Neuroreport* 2: 485–488.

Priebe, N. J., & Ferster, D. (2005). Direction selectivity of excitation and inhibition in simple cells of the cat primary visual cortex. *Neuron* 45: 133–145.

Rancz, E. A., Franks, K. M., Schwarz, M. K., Pichler, B., Schaefer, A. T., & Margrie, T. W. (2011). Transfection via whole-cell recording in vivo: bridging single-cell physiology, genetics and connectomics. *Nat Neurosci* 14: 527–532.

Rancz, E. A., Ishikawa, T., Duguid, I., Chadderton, P., Mahon, S., & Hausser, M. (2007). High-fidelity transmission of sensory information by single cerebellar mossy fibre boutons. *Nature* 450: 1245–1248.

Stuart, D. G., & Brownstone, R. M. (2011). The beginning of intracellular recording in spinal neurons: facts, reflections, and speculations. *Brain Res* 1409: 62–92.

Tan, A. Y., Brown, B. D., Scholl, B., Mohanty, D., & Priebe, N. J. (2011). Orientation selectivity of synaptic input to neurons in mouse and cat primary visual cortex. *J Neurosci* 31: 12339–12350.

Tan, A. Y., Chen, Y., Scholl, B., Seidemann, E., & Priebe, N. J. (2014) Sensory stimulation shifts visual cortex from synchronous to asynchronous states. *Nature* 509: 226–229.

Wang, X., Vaingankar, V., Sanchez, C. S., Sommer, F. T., & Hirsch, J. A. (2011). Thalamic interneurons and relay cells use complementary synaptic mechanisms for visual processing. *Nat Neurosci* 14: 224–231.

Wehr, M., & Zador, A. M. (2003). Balanced inhibition underlies tuning and sharpens spike timing in auditory cortex. *Nature* 426: 442–446.

Woodbury, J. W., & Patton, H. D. (1952). Electrical activity of single spinal cord elements. *Cold Spring Harb Symp Quant Biol* 17: 185–188.

Zhang, L. I., Tan, A. Y., Schreiner, C. E., & Merzenich, M. M. (2003). Topography and synaptic shaping of direction selectivity in primary auditory cortex. *Nature* 424: 201–205.

Extracellular Single-Unit Recording and Neuropharmacological Methods

William L. Coleman and R. Michael Burger

Introduction

Proper cell to cell communication is crucial for multicellular organisms to maintain homeostasis, to generate behaviors, and ultimately to survive. In some cases neural signals need to rapidly and efficiently travel long distances in order to mediate these effects. The primary information-carrying neural signal is the action potential or "spike." When neurons fire action potentials, the movement of ions across the cell membrane causes a transient voltage change in the range of millivolts. These tiny events can be detected and monitored over long periods without disrupting the cell using electrodes specialized for extracellular recording. Electrodes can be tailored in such a way that the electrical activity of individual neurons very close to the tip of the electrode can be isolated from other sources of voltage noise. This recording arrangement is fundamental to extracellular single-unit recordings of both vertebrate and invertebrate neurons in situ. This chapter will focus on general aspects of this technique and the equipment used. In addition, we will give tips on some strategies to optimize recording efficiency. We will also highlight some specific cases from our experiments on auditory neurons in birds and mammals to demonstrate some of these principles in a practical context. While these examples are specific to our field of study, the basic principles of extracellular recording that will be discussed can be applied to many different experimental preparations, including invertebrate animals.

As with any technique, *in vivo* extracellular single-unit recording has its advantages and disadvantages. The real power of *in vivo* single-unit recording is the ability to measure physiological response properties that reveal how individual neurons process and represent information in their spike patterns. These information-bearing properties may include spike threshold, latency to response, firing rates (both spontaneous and driven), dynamic range, and temporal firing patterns. Cells with similar response properties can be categorized into groups, which perhaps represent different physiological or computational functions. These physiologically defined groups of neurons are often correlated with anatomically grouped neurons in spatial location, morphology, or both. Another advantage of extracellular recording is that it does not require the penetration or disruption of the neuronal membrane. Membrane damage occurs by default in intracellular recordings, which can be problematic as neuronal viability and physiology are often compromised. Although intracellular recordings have the advantage of detecting the full millivolt range of postsynaptic potentials, extracellular recordings are still quite capable of resolving the local microvolt-range signals available from outside the neuron. Neuronal firing patterns are best measured when stress or damage to the cell's integrity is minimized, and in many cases an individual neuron can be recorded from for tens of minutes to hours.

A powerful related technique that we will discuss at length is the use of "piggyback"-style multibarrel electrodes, which allow for thorough characterization of physiological responses during focal application of pharmacological agents. The latter allows for the experimental manipulation of individual classes of synaptic receptors, or other presynaptic and postsynaptic proteins in the local milieu. Finally, extracellular recording techniques often allow for multiple neurons to be recorded from within the same tissue penetration.

Of course, both multibarrel and single-barrel extracellular recording methods include caveats and pitfalls unique to *in vivo* recording. These techniques necessitate the use of a living animal subject. Animal preparations can involve difficult surgery and can require complex anesthesia protocols, which may influence neural properties. Additionally, an apparatus that maintains the animal as securely and comfortably as possible while optimizing conditions for stable recordings is required. For example, minute sources of motion such as those conferred by respiration can present a challenge for maintaining good-quality single-neuron isolation for long periods. In experiments that require awake and/or freely moving subjects, stable isolation of individual neurons may be extremely difficult and require a more elaborate apparatus for electrode implantation (Chapter 4). Another challenge of extracellular recording is locating the targeted neurons in the absence of visual guidance, especially in deep brain structures. This usually necessitates stereotaxic methods of electrode positioning and the need to verify the location of the recording site post hoc. The latter can be achieved with various histological techniques following the recording session. A few specific examples of protocols will be discussed in a later section. Finally, while extracellular recording offers the experimenter the advantage of monitoring neural output with relative ease, the technique generally does not provide a nuanced view of the myriad synaptic or biophysical factors that contribute to a neuron's responses. For this information whole-cell patch clamp techniques are required. These techniques are reviewed in Chapters 1 and 2.

The ultimate practical impact of these advantages and disadvantages depends on the particular experimental system. For example, invertebrates tend to have fewer, larger neurons that can often be visually or physiologically identified from organism to organism. Fundamental to all preparations is the need to electrically isolate individual neurons and then maintain a stable recording long enough to complete the experimental protocols. The basic methodology of extracellular single unit recordings, including the general recording schematic, necessary hardware and equipment, and necessary software will next be described in detail, with specific examples provided from *in vivo* recordings of chicken auditory brainstem neurons.

Methodology

General Extracellular Recording Setup

While specialized equipment may be necessary to accomplish individual experimental goals, any setup for extracellular recordings at its core will include the following key components: (1) a recording electrode connected to a high-impedance headstage, (2) a reference/ground wire, (3) a differential amplifier, (4) a filter, and (5) some means of monitoring the amplifier output, such as an oscilloscope and/or computer with recording software. If a computer is used, an analog–digital converter will also be required. An example of a basic extracellular recording setup is shown in Figure 3.1. There are many ways that these core components can be incorporated into a recording setup to suit either experimental needs or personal preference. In some cases key components can be combined—for example, many amplifiers include some filtering circuitry, but separate filtering units can also be used. Keeping this basic setup in mind, specific examples of *in vivo* extracellular single unit recordings will be described in detail.

Recording Electrode Preparation and Control

Recording electrodes are typically fabricated from either glass or metal. In the first case, glass capillary tubes are heated until melted and "pulled" until they separate, leaving a fine tip at both ends. Several "pipette pullers" are commercially available, and they typically apply both heat and force, via magnets or gravity, where settings for each can be adjusted to produce tips of different shapes, sizes, and resistances. Glass electrodes with long and thin tapering tips are well suited for *in vivo* neuron recordings as they more readily pass through tissue with less damage and reduce problems that may arise due to size constraints. For example, difficulties arise when attempting to pass a large electrode through a small craniotomy. The pulled capillaries are then filled with a conductive electrolyte solution. Solutions commonly used to fill electrodes include near-saturated KCl solution (3 M) or physiological saline. Electrodes can be filled by using commercially available microfillers that attach to syringes, but

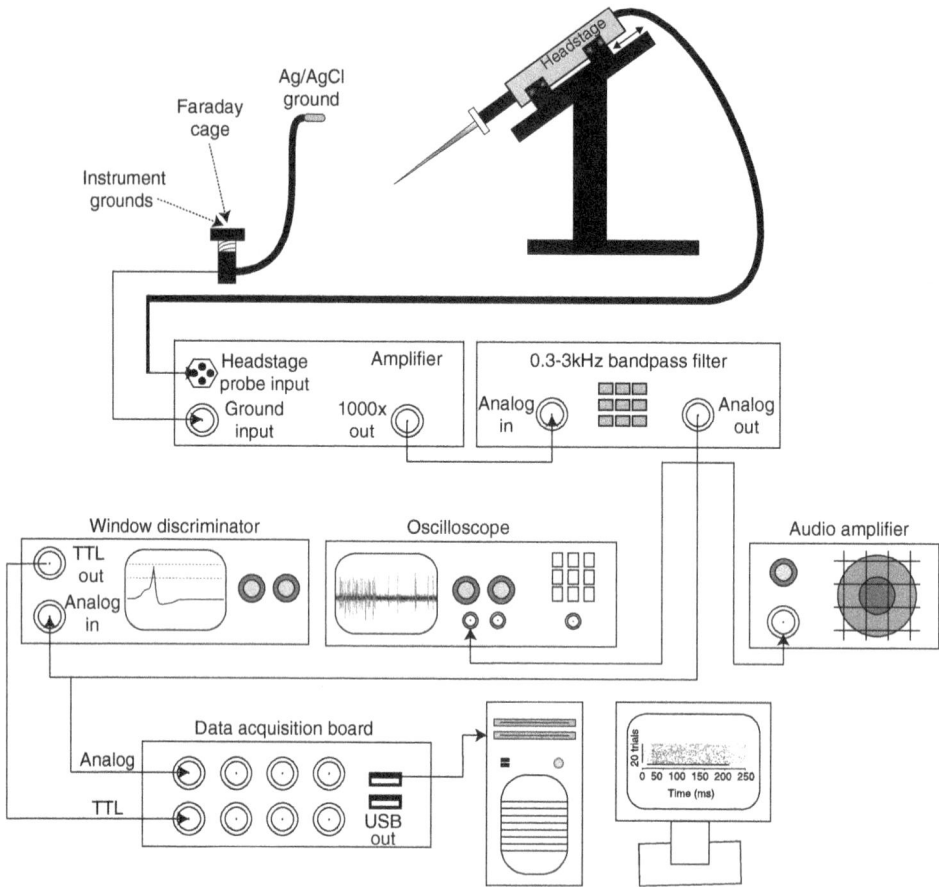

FIGURE 3.1 A basic extracellular electrophysiology setup including essential equipment. The electrode and headstage are mounted on a movable actuator that may be motorized for remote electrode control. The ground wire should be placed in liquid contact with the preparation (not shown). All instruments, the Faraday cage, and sources of noise should be grounded to a central point, represented here as a metal screw on the vibration table, prior to amplification. The ground and headstage provide input to the differential amplifier, which will generate an output signal ranging from 10^0 to 10^4x. This signal is sent to a filter. A band-pass filter set from 300 to 3,000 Hz is ideal for extracellular recording. The amplified and filtered signal can then be monitored on an oscilloscope and/or audio monitor. Most modern rigs are set up to record this signal using analog–digital conversion on a data-acquisition board. Once digitized, this signal can be recorded in a computer file. In addition, or as an alternative, a window discriminator that sends a short-duration TTL pulse to the computer when spikes are detected can report action potential event times. In this case, only spike times are recorded and the waveform is lost. These two recording methods can be combined to take advantage of both kinds of data.

these are expensive and tend to break easily. Handmade electrode fillers can be constructed by carefully heating a plastic tube (either a disposable transfer pipette or plastic micropipetter tip) over a small flame and pulling until the desired thinness is acquired. This technique takes practice and skill but when done correctly works as well as the commercially available microfillers. Pipettes are then electrically coupled to the headstage. This can be achieved in several ways, again depending on the constraints or needs of the experimental setup. The glass capillaries can be inserted directly into various types of electrode holders that have either a silver wire that runs inside the glass or a silver/silver chloride pellet. Silver/silver chloride electrodes are typically used because they are reversible and nonpolarizing; this means that as current flows, other ions cannot build up on the silver surface over time, and thus electrode potentials remain stable. This so-called junction potential occurs whenever two conductors of differing composition come into contact. Junction potentials can introduce errors into recordings or even prevent the amplifier from operating properly. Use of silver/silver chloride conductors requires that the electrolyte solution in the pipette contain Cl⁻. Electrode holders have a connector pin that plugs into the headstage. Electrodes can be coupled to the headstage at a distance by using silver/silver chloride wire attachments, which can be fabricated in the laboratory. Teflon-coated silver wire can be soldered to a connector pin on one end, and on the other end a small amount of Teflon can be gently removed with a razor blade and the exposed silver wire can be chlorided, either by soaking in chlorine bleach followed by thorough rinsing, or by placing in KCl solution and attaching a voltage source (silver wire to anode). Silver/silver chloride wire will appear as dull or darkened compared to pure silver.

Electrode Properties and Configurations

Glass electrodes are highly malleable when heated and can be bent to different shapes and fabricated in many configurations to suit experimental needs. One experimentally useful example is the construction of "piggyback"-style multibarrel electrodes (Havey and Caspary 1980), which will be discussed in detail in a later section. A major advantage of recording with glass pipettes is that these electrodes can also be filled with solutions that contain anatomical tracers or other markers that can be delivered immediately following the recording.

A major disadvantage of glass electrodes is that they are fragile, so they can generally be used during only a single experiment, and each experiment may require several pipettes. This may become a particularly aggravating factor when using piggyback multibarrel electrodes, which require considerable time and skill to prepare and cannot be fabricated on demand during experiments. An alternative is metal or carbon-based electrodes, which can be fabricated in the laboratory but are also commercially available in a wide variety of sizes and resistances. These typically involve an insulated metal wire that is exposed at the tip. A connector on the other end attaches to the headstage, similar to glass electrodes. Metal electrodes do not have the fragility of glass, and the same electrode may be reused day after day over multiple experiments. Recording sites can be marked following experiments with electrolytic lesions by passing current through the metal electrode. However, excessive current may melt the tip of the electrode, lower its resistance, and make it unusable for isolating individual neurons. Metal electrodes are also more difficult to use for fabrication of custom electrodes, such as piggyback electrodes.

Regardless of the electrode material, movement and position are typically controlled by a micromanipulator that has mechanical, hydraulic fluid, or piezoelectrical controls. These can attach to the headstage directly or to an electrode holder that connects to the headstage at a distance. In some cases, an actuator can be used in addition to a micromanipulator to control electrode position. A myriad of commercially available or custom-made devices are available to meet the specifications of a specific experimental preparation or stereotaxic device.

Preparing for an *in Vivo* Recording Experiment

Extracellular recording from isolated neurons *in vivo* usually requires some dissection or surgical preparation. Careful thought must be given to how organisms will be treated throughout the experiments. Of foremost concern is the minimization of pain or discomfort to the subjects. Use of vertebrate organisms will require prior approval by an Institutional Animal Care and Use Committee (IACUC) or similar governing committee. It is useful to have a designated dissection and surgery area that has space for everything that may be required. While the actual dissection and surgical tools needed may vary depending on specific experimental needs, some basic tools may include scissors, forceps, hemostats, cauterizer, styptic powder or other antihemorrhagic agent, sterile gauze, surgical Gelfoam, surgical suturing, and so forth. A proper light source, such as a bifurcated fiber-optic illuminator, is also very useful. Many types of dissecting microscopes are available, which can aid in the visualization of fine structures.

Prior to experimentation, organisms must be anesthetized in some way to ensure a stable condition necessary for proper surgery or dissection. Anesthesia must be properly maintained during experiments so that the organisms do not feel pain or discomfort, and in cases where animals are required to recover from anesthesia, analgesics must be used. The complexity of anesthesia varies depending on the organism to be used. In the case of invertebrates, anesthesia can be as simple as placing the organism in ice; organisms where chemicals can readily cross the integument (e.g., worms, amphibians, or fish) can be soaked in a water-soluble anesthetic or analgesic prior to dissection. Vertebrate preparations are more complex and the anesthetic agents used depend on the specific experimental goals. Inhalant anesthetics such as isoflurane or halothane can be used, but these typically require some type of vaporizer system that may not be practical for use on small organisms. Another option is to use pharmaceutical agents that can be injected into muscle, subcutaneously, or into the peritoneal cavity. The purchase of some chemicals may require regulatory approval and secure storage conditions. Anesthetic agents must be chosen carefully and with consideration of the physiological site of action, especially with regard to effects on neuronal properties. An example of a typical *in vivo* single-unit recording scenario will be described in the next section with hints along the way to aid in recording optimization. Data-recording format and analysis will also be discussed briefly.

Anesthesia and Surgical Preparation

The method of anesthesia is a critical element of experimental design for extracellular physiology. Careful consideration of various anesthesia issues must be addressed prior to embarking on a new experiment. First, it is important to consider the site of action for the chosen anesthetic agent (e.g., which receptors and ion channels are affected). This will determine whether the use of a particular drug might confound the results. For example, one would not choose barbiturates if the experiment were designed to test the role of GABA in neural processing. Choice of an anesthetic also depends on how long deep anesthesia must be maintained. Consideration of these questions and more should be resolved in consultation with a qualified veterinarian prior to IACUC approval. In addition to general anesthesia, topical analgesics at the surgery site may be useful to suppress painful stimuli following or during the experiment.

Thermal homeostasis is normally disrupted during deep anesthesia. Maintenance of normal body temperature will preserve normal physiological response properties (e.g., thresholds) and help sustain subject viability for a longer period. Temperature can be monitored by a probe inserted into the rectum/cloaca. This signal can then be coupled to a feedback heating pad under or surrounding the animal to externally maintain body temperature within the normal range.

Surgery

The exact surgical approach will, of course, depend on the demands of the experiment. Here we describe a typical procedure for recording from the brain of a small rodent or bird. First, depilatory cream can be used to clear the skin of hair or feathers prior to making an incision. Second, a midline incision can be made exposing the muscles on the dorsal surface of the skull. A cautery or scraping device is used to reflect the muscle tissue from the surface to expose and dry the bone so that adhesive will stick. Third, a rigid system to fix the head in place can be mounted on the skull using cyanoacrylate glue and/or dental acrylic. In our case, we use a brass pin, which can be rigidly clamped to the stereotaxic apparatus. The stereotaxic apparatus and all recording equipment are attached to a level floating antivibration table.

At this point the animal should be secured in a comfortable holder and the temperature-control equipment should be in place. A craniotomy is then performed at the appropriate coordinates with a dental drill. Coordinates are often calculated from skull suture landmarks like lambda and/or bregma. Care should be taken to remove all bone from the opening, and the borders of the craniotomy should be checked with fine forceps or a blunt probe to rule out the presence of difficult-to-see bone flakes remaining from the procedure. Additionally, the ground wire lead can be placed either into the muscle tissue or under the skin. Some researchers prefer to drill a second craniotomy and place the low-impedance reference electrode directly into the brain. In any case, the ground wire must be in liquid contact with the animal to ensure good signal-to-noise quality. A key point is to have all metal surfaces (e.g., Faraday cage, stereotaxic, light sources) and grounds for all instruments coupled to a common point in the apparatus that is itself coupled to the ground input of the amplifier. This strategy effectively reduces unwanted interference from ground loops or other sources of noise. A sharp tungsten probe can be used to tear and reflect the dura mater from the opening to ease electrode penetrations. Medical-grade silicone gel can then be used to fill the craniotomy to prevent drying. This latter step generally will not degrade electrode quality if the layer of silicone is kept relatively thin.

Next, the electrode should be properly positioned with respect to the skull's orientation and advanced to the surface of the skull. Inclusion of dye such as 4% fast green in the pipette solution can aid visualization of fine glass electrodes at the brain surface without compromising physiological responses. Finally, controlled advancement of the electrode can commence with a micromanipulator or ideally with a motorized micropositioner–actuator.

Locating the Target Area

Locating the area of interest can be challenging, particularly when it is not a superficial structure and readily visible. There are several ways to improve the probability of quickly and consistently finding the target area, which maximizes time for experimentation and data collection. The first is consistency in the experimental setup. This can be accomplished by minimizing variation among organisms (using animals of similar age and size), making sure that all equipment stays level (the floating table, the platform, the electrode holder), and ensuring that the subject's orientation and spatial position (degree of roll, pitch, and yaw) are as similar as possible from experiment to experiment. Careful study of the anatomy of the target area can also be very helpful prior to the experiment. Electrode position should be verified post hoc using histological techniques. Generally, this involves aldehyde fixation of the tissue, creating thin slices with a microtome, and using a histological stain to visualize cells. Not only can this technique be used to study the anatomy surrounding the target area, but in some cases "electrode tracks," or areas of physical damage caused by passing electrodes through the tissue, can be visualized to improve subsequent electrode placement. Labeling substances can be ejected from the electrode to confirm the recording site, which will be discussed in detail in the next section.

We often used a blunted (low-impedance) electrode during initial penetrations to physiologically confirm the recording site based on typical "background" activity, which is more detectable with blunt electrodes (0.5–1 MΩ). Glass electrodes in this resistance range are generally capable of detecting field responses from multiple neurons but typically are not able to isolate individual neurons. Once confident of the recording depth and position, we discard the blunt electrode for a sharp one to isolate single-neuron responses.

Isolating and Recording from Individual Neurons

Sharp electrodes with an electrical resistance range of 2–15 MΩ or higher are ideal for isolation of individual neurons. At this point, the sharp electrode (or piggyback multibarrel) is positioned using the same reference marks and then driven to the same depth coordinate range where the target area was previously encountered with the blunt electrode. The electrode penetration ideally proceeds very slowly (in the micrometer per second range) through the tissue to avoid breakage of the tip. The resistance of the pipette should be checked frequently with the amplifier's test circuit. Electrode blockage is usually manifested as a sudden change in the background noise detected on the audio amplifier and confirmed by a drastic increase in resistance during the "electrode test." The electrode test typically passes a known current through the electrode, and the resulting voltage deflection indicates electrode resistance according to Ohm's law. One can attempt to clear the blockage by turning on continuous current injection and/ or switching the polarity of the current injection back and forth several times. Another helpful way to clean a blocked electrode is to use the "capacitance override" or "zap" button if the amplifier is equipped with one. This feature is part of the built-in capacity controls of the amplifier, which act as a feedback loop for counterbalancing the capacitance of recording electrodes. Electrode capacitance effectively low-pass-filters neural signals and "rounds off" waveforms. This problem can be compensated for by adjusting the capacity compensation knobs. Overcompensation produces feedback oscillations or "ringing," and signals cannot be recorded in this condition. However, in the "ringing" state, the rapidly oscillating signal causes the electrode to physically vibrate, which can sometimes unblock the tip of the electrode. In some cases this oscillation property can assist in clearing a clogged electrode. While advancing the electrode through the tissue, it is sometimes necessary to clean the electrode frequently using the above methods, particularly when within the target area. If the electrode is blocked and the resistance is too high, the small biogenic voltage signals cannot be detected. If at any point the electrode becomes irreversibly blocked and cannot be cleaned, it must be replaced.

Another issue often encountered during experiments is drifting baseline voltage, the steady-state voltage that generally is considered "ground." The baseline voltage level can be adjusted using the "DC offset" knob on the amplifier, which can counterbalance or "offset" baseline electrode potentials, which can vary during an experiment. It is possible for the baseline voltage to drive the amplifier "out of range," meaning that it is beyond the operating limits of the amplifier. In this situation, amplifier output will flat-line and time-varying electrical signals cannot be detected. Adjustment of the "DC offset" will typically solve this problem. If this problem occurs frequently, the electrode should be replaced, and both the probe and reference electrode should be checked for adequate silver/silver chloride coating. Removal of this coating is the most common cause for drift in the reference voltage and DC offsets that exceed the amplifier's operational range.

Another important consideration prior to recording biogenic signals is setting a filter to reduce or eliminate noise. Some amplifiers have built-in "notch" filters that can remove signals with a specific frequency, such as typical line noise frequencies of 50 or 60 Hz. Additional filter units can also be used. Extracellular action potential signals are typically band-pass–filtered between 300 Hz and 3 kHz to preserve the action potential waveform but also to eliminate unwanted noise.

Once the target area has been reached, as evidenced by stereotaxic position or other forms of feedback to the experimenter such as characteristic background activity, the goal is then to isolate and record from individual neurons, or "units." The electrode should be advanced very slowly, with frequent pauses between forward movements. This allows for "settling" of the tissue around the electrode, which improves recording stability. As the electrode draws closer to an individual neuron, distinct "pops" can be heard above the background noise with a speaker coupled to the amplifier output, and waveforms that are larger than the baseline noise become visible. At this point, it is prudent to wait and let the tissue settle rather than keep advancing the electrode. In some cases, the neuron will be "lost" and the searching must recommence. Other times the signal-to-noise quality will dramatically increase, and clear action potential waveforms will be discernible. In this case, one should stop advancing the electrode and should perhaps retreat a few microns. Occasionally, advancement toward an already isolated unit will irritate or even kill the neuron. These neurons are typically "lost," and, in any case of apparent recovery, responses would not be representative of normal physiological conditions. After each encounter with an isolated neuron, clearing the electrode with current or the capacitance override is advised before attempting to isolate a new neuron. In some cases, individual neurons may be isolated just after cleaning the electrode. Waveforms from isolated neurons have certain characteristics that are distinct from multiunit recordings. The signal-to-noise ratio will typically be large enough that the waveforms can be easily distinguished from the background noise. Waveforms will be relatively uniform in shape, amplitude, and duration. The presence of multiple peaks, inconsistent waveforms, or spikes that are very clearly different in height are indications that neurons are not well isolated. The voltage window discriminator, when triggered by a spike, sends transistor-transistor logic (TTL) pulses to the processor. These TTL pulses represent action potential event times and can be set to detect one spike waveform or another. For example, if waveforms of two clearly different amplitudes are observed, one can set a detection window (above, below, or within) based on peak voltage range, which excludes detection of spikes that do not fit the desired criteria.

Prior to actually recording from isolated neurons, it is important to consider what stimulus parameters will be used and how the data will be recorded. Depending on the experimental equipment used, stimulus parameters are best set ahead of time and either saved to files that can be quickly loaded using software or stored directly on the stimulus hardware. Recording duration is variable and unpredictable, and therefore having a prepared set of stimuli on hand is beneficial to maximize data collection per neuron. Two main types of data can be collected during extracellular recordings: voltage waveforms and event times, as shown in Figure 3.2. Voltage waveforms can be used to analyze spike frequency and firing patterns and general waveform shape. One must be cautious when attempting to interpret transient properties of an extracellularly recorded waveform, which are highly dependent on the position of the recording electrode in relation to the neuron. For example, if the recording electrode and neuron become further apart the

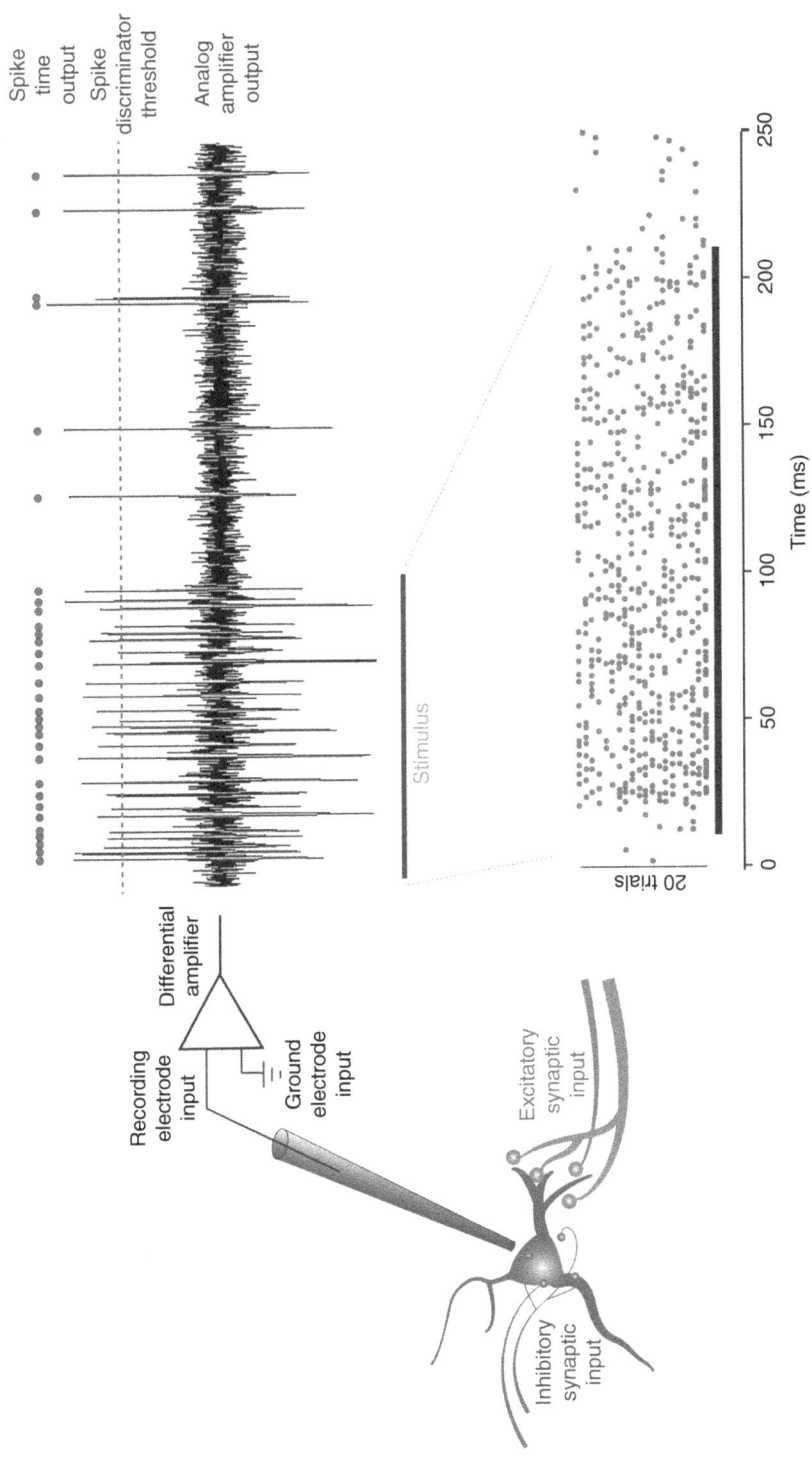

FIGURE 3.2 Schematic representation of the extracellular configuration and data formats. Left: A glass pipette filled with electrolyte and threaded with an AgCl-coated silver wire approaches a neuron. The amplifier inputs monitor the voltage at the open electrode tip and the extracellular space. Top right: Action potentials are isolated from the noisy background when the electrode approaches close enough to the target neuron. Spikes can be detected with a window discriminator, which is set to trigger at a particular threshold, shown with the upper dashed line. Bottom right: Spike times can then be used to build histograms (not shown) or raster displays of response characteristics over several stimulus trials.

spike amplitude will decrease. It is ambiguous as to whether such changes are due to physiological phenomena or simply the electrode position.

Digital recording of the voltage waveforms also may require more processing power to accommodate a sufficiently high sampling rate, and the resulting files are large. Recording event times, in contrast, also allows for analysis of spike frequency and firing patterns but requires less processing power and storage space. Spike time data can be displayed in such a way that changes in firing pattern are easily visualized. Event time data are often displayed in raster plots, where a "dot" or other marker represents the occurrence of a spike at a particular time.

The Piggyback Electrode for Pharmacological Manipulation

After mastering the basics of extracellular recordings from isolated neurons with single sharp electrodes, "piggyback"-style multibarrel electrodes can be used to investigate the physiological contributions of many different targets, including ion channels, receptors, and/or other molecules of interest. Single-barrel glass capillaries pulled to a fine tip can be manually bent over a small flame and then glued to a separate multibarrel capillary glass that has also been pulled and manually broken back to a smooth edge. The protruding sharp recording electrode can be used to isolate neurons using the same methods as with a simple single electrode. This arrangement has the added benefit of manipulating the pharmacological environment very locally in the recorded neuron's environment. Drugs that, for example, block specific neurotransmitters can be applied through the multibarrel electrode via iontophoresis (see Fig. 3.3, and Coleman et al. 2011). Two to five barrels containing drugs can be used, but in each case, one barrel should be reserved to provide a current equal and opposite in sign to the sum of currents applied to the drug barrels. This feature is usually provided in iontophoresis systems and is necessary to avoid electrically influencing the recorded neuron. Each barrel is connected to a headstage with silver/silver chloride wires and each barrel is controlled by an iontophoresis module or "channel." We use this setup to investigate the contributions of inhibitory neurotransmitter systems (both GABAergic and glycinergic) in shaping the physiological responses of auditory neurons to acoustic stimuli, but there are many conceivable applications of this method. As a brief example, drugs can be given together as a "cocktail" or sequentially. Bundled capillaries with three, five, and seven barrels are commercially available. Additionally, while we use receptor antagonists, it is also possible to apply agonists, protein inhibitors, or other chemicals using this method. Fabrication of piggyback multibarrel electrodes is shown in Figure 3.3.

Using piggyback multibarrels does have some disadvantages. As discussed earlier in this chapter, they are difficult and time-consuming to make properly. The tip of the multibarrel itself is wider than the single sharp recording electrode and is more difficult to pass through tissue, and the higher the number of barrels used, the more problematic

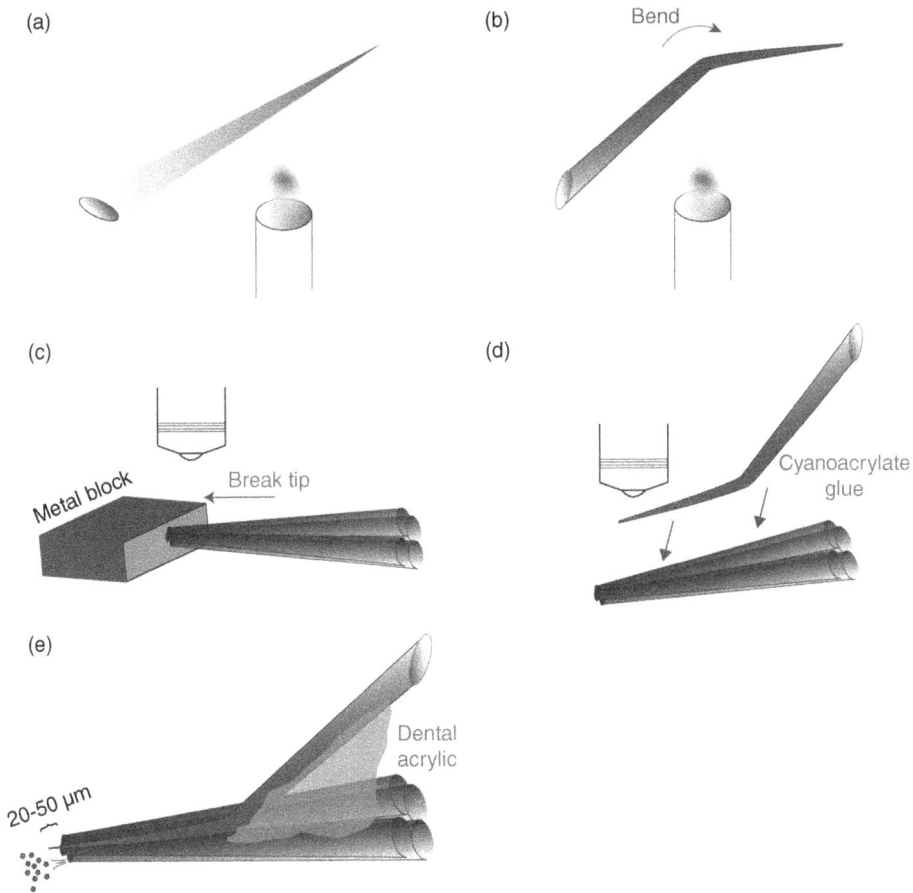

FIGURE 3.3 Fabrication of piggyback multibarrel electrodes. First, a single-barrel capillary glass recording electrode and multibarrel blank electrode are each pulled to a sharp tip (not shown). A, B: The recording barrel is then heated over a very small flame and bent to accommodate gluing the tapered tip to the multibarrel. C: The sharp multibarrel is gently moved into a metal or plastic block until a blunted and squared-off 50- to 80-µm tip width is achieved. Each component barrel would then have about a 10- to 30-µm diameter. Careful monitoring of this process with a microscope ensures quality. D: Again under the microscope, and with the use of a micromanipulator (not shown), lower the bent recording electrode until flush with the multibarrel. Allow a 20- to 50-µm protrusion of the sharp electrode at the tip. It is essential to assess good alignment of the tips prior to gluing. A gentle tap on the table will confirm whether or not the tips are well coupled to each other. A small amount of cyanoacrylate glue is applied at this stage to fuse the two electrodes. It is essential to use no more glue than needed, as it can easily wick into the barrels and block iontophoresis. E: Once the glue is completely dry, reinforcement of this bond is achieved with careful application of dental acrylic. Once sufficient time for the acrylic to cure has passed, the electrode tip should be viewed under the microscope again from both the top and the side to ensure that tip separation has not occurred.

this becomes. Electrode width increases rapidly extending away from the tip and can exacerbate tissue damage or reach the borders of a small craniotomy more readily than with a single-barrel configuration. Thus, accommodations for tissue damage or electrode trajectory have to be made when recording from neurons that are deep in the tissue. For example, a larger-diameter craniotomy may be necessary to accommodate the bulky "back end" of the piggyback electrode. The multibarrels are prone to blockage from either tissue or cyanoacrylic glue during fabrication, and this blockage usually will not be detected before use. Aside from these issues, actually isolating and recording from individual neurons is much the same as described for single sharp electrodes. Charged drug compounds can be retained in the barrels with current of opposite charge to that of the ionized drugs (control) and then expelled by switching the polarity of the current (treatment), typically until a saturated effect is observed. The drug ejection can then be turned off, and the neuron can return to predrug conditions (recovery). In many cases, neurons can be held long enough to test multiple drugs and allow for recovery. Methods of confirming the recording site will now be discussed.

Confirmation of the Recording Site

A common difficulty with *in vivo* recording is that often the target area is deep in tissue and is not readily visualized. This necessitates post hoc confirmation of recording sites. There are several ways to achieve this, depending on the electrode material. For example, certain anatomical tracers can be included directly in glass recording electrodes, and the tracers can then be ejected following experiments by using current injection, provided the tracer molecule has a known charge. Good options include fluorescently labeled compounds that are taken up by living neurons (e.g. rhodamine-conjugated dextrans) or a compound that can be developed with a precipitate that forms a dark reaction product when processed (e.g., neurobiotin, developed with avidin–biotin–peroxidase reaction and diaminobenzidine [DAB]). Different tracers may necessitate the use of different electrolyte solutions in the pipettes. For example, Köppl and Carr (2008) dissolved neurobiotin in 2M K acetate rather than 3M KCl. For metal electrodes, current injection can also be used, but in this case a lesion or "burn mark" in the tissue will be created. In each case, the localized label can be used in conjunction with careful systematic electrode placement to track the location of recording areas in the tissue. Current injection can be controlled either by the amplifier or by carefully disconnecting the silver/silver chloride wire from the headstage, to avoid damage to its circuitry, and attaching the wire to an external iontophoresis unit. One should wait at least 30 minutes to 1 hour following either the loading of a tracer to allow proper uptake into cells or to allow for an inflammatory response following the lesion, which will make it more visible in the tissue.

Following this period, the tissue must be fixed and processed according to a protocol that is appropriate for the specific label used (e.g., there are different protocols and requirements for a fluorescent markers vs. markers that depend on additional steps to generate a visible reaction product). In general, a basic protocol includes the following steps. Animals are deeply anesthetized prior to transcardial perfusion first with phosphate-buffered saline (PBS) that may include anticoagulants like heparin, followed by 4% paraformaldehyde in PBS (this is a typical perfusion protocol that can used for many organisms). Following fixation, the brain is carefully removed from the skull and may be postfixed overnight in 4% paraformaldehyde at 4°C. Following thorough rinsing in PBS, the brain tissue of interest is blocked and sliced on a microtome. If a freezing microtome or cryostat microtome is to be used, an additional cryoprotection step involving submersion of the tissue in a 30% sucrose solution overnight is necessary to draw water out of the tissue. Ice formation within cells will destroy the histological quality of the preparation. Section thickness depends on the application but can range from 10 to 200 microns. Finer sections are ideal for detailed histological analysis while thick sections may be sufficient to verify electrode placement. Slices containing the structures of interest are mounted on a gelatin-coated glass slide in a medium that is compatible with the tissue preparation, and coverslipped. Recording-site labels or marks are visualized using an appropriate microscopy method (e.g., confocal, epifluorescent, or light microscopy). Some specific nuances to enhance images for each type of labeling protocol will now be discussed.

Fluorescent dextrans. Neurons and glia can be counterstained with fluorescently tagged Nissl stain (different colors are available; green/red contrast between tracer and Nissl marker would be a typical choice). Fluorescent labeled slides should be coverslipped using an aqueous mounting medium that helps prevent photobleaching, such as Glycergel (Dako) or Vectashield (Vector Laboratories).

Biotin/Neurobiotin. Biotinylated compounds can be developed using the Vectastain ABC kit (Vector Laboratories) and coverslipped in a nonaqueous medium such as Permount (Fisher Scientific). An example of a neurobiotin-labeled recording site is shown in Figure 3.4 (see also Köppl and Carr 2008, and Coleman et al. 2011).

Lesions. Some counterstain should be used to help more precisely identify the lesion site. Fluorescent Nissl stain can be used but is not necessary. Cresyl violet or toluidine blue staining can be used to visualize neurons and glia for light microscopy. These slides should also be coverslipped in Permount (or a similar mounting medium). (For an example of using metal electrodes to record from auditory neurons and using lesions to mark recording sites, see Tabor et al. 2012.) Some general examples of data analysis will be discussed in the next section.

Data Analysis

The type of data analysis required depends on the data output recorded (analog waveforms, spike times, or both) and what the specific experimental needs are. Various software packages and programs for data analysis are commercially available. These often have many features but can be rather expensive and limited to use with specific equipment or other compatible software. One example of packaged software that can be used to analyze waveforms is pClamp, which includes Axoscope, Clampex (used for data acquisition), and Clampfit (for analysis). These programs specifically require the use of an Axon Digidata digitizer (all from Molecular Devices). It is possible to create your own analysis tools using MATLAB (MathWorks), Igor (Wavemetrics), or more general programming languages. Many research laboratories also make their custom software available to the public free of charge. How data will be analyzed must be carefully

(a)

(b)

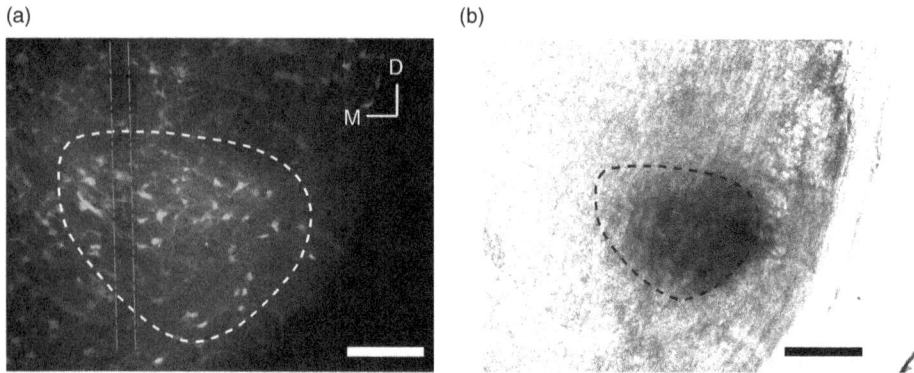

FIGURE 3.4 Histological markers for recording site verification. A: Panel shows neurons labeled near the recording site with rhodamine-conjugated dextran injected at the end of the experiment. The structure is the superior olivary nucleus in the chicken auditory pathway. Parallel near vertical lines show the electrode path through the tissue. Scale bar = 200 µm. B: A similar injection of neurobiotin into another superior olivary nucleus at the recording site is developed with a peroxidase reaction with SG (Vector Laboratories) to give a dark reaction product. This method is more archival than fluorescent labels.

considered along with any possible compatibility issues (or specific requirements) prior to purchasing expensive equipment or software packages.

Equipment and Reagents

In many cases, the equipment required for extracellular single-unit recordings may vary based on experimental needs or personal preference, and there may be many other options beyond what is listed below. The following list is based only on our previous experience but will provide examples of what may be required and may be a benefit to researchers as a starting point to explore purchasing equipment for their own needs.

Electrodes

- *Metal electrodes.* These are typically insulated metal wires that can have different diameters, tip tapers, and electrical resistances. For example, epoxy-insulated tungsten or stainless-steel microelectrodes are available (A-M Systems, Frederic Haer, World Precision Instruments).
- *Borosilicate glass capillary tubes.* Different internal and external diameters are available, with or without internal filaments, single or multibarrel (A-M Systems, or World Precision Instruments, Harvard Apparatus).
- *Pipette puller.* A wide range of pullers is available, each with its own strengths and weaknesses. Horizontal pullers, such as the P-97 (Sutter Instrument Co.) or PMP-107 (World Precision Instruments), are programmable and different settings can be stored. There are also vertical pullers, like the PC-10 for pulling single-barrel electrodes and

the PE-22 for pulling multibarrels (both from Narishige). Many pullers can be set to have single- or multiple-stage pulls to create optimal electrode characteristics for the given application.

- *Pipette fillers*. MicroFil nonmetallic syringes (World Precision Instruments). A less expensive (but less reliable) alternative can also be handmade by heating and carefully pulling plastic labware, like plastic pipette tips, 1-mL syringes, or disposable transfer pipettes.
- *Syringe filters*. Used to filter solutions prior to filling pipettes; different pore sizes are available (Millipore, WPI, or Fisher Scientific).
- *Syringes*. Used to fill electrodes; many different volumes are available.
- *Metal wire*. Silver wire is often chlorided and used to create electrodes or connect electrodes to headstages, but other metals are also available. Wire is available in different diameters and can be insulated or bare (A-M Systems or World Precision Instruments).
- *Miscellaneous electrical items*. Metal pin connectors (male and female), BNC connectors, alligator clips, insulated wire, and so forth. These are available from A-M Systems and World Precision Instruments but can often be purchased at a local electronics store.
- *Soldering iron/solder*. Can be used to create a physically secure electrically continuous connection between two metals.
- *Electrode holders*. These are typically plastic, ceramic, or metal cylinders, which may have a screw and/or gasket to hold the electrode in position and couple the electrode to the manipulator or motorized actuator. There are many shapes, sizes, and configurations (e.g., some have a port for applying pressure or suction). They may have a silver wire that runs inside the recording electrode or a silver/silver chloride pellet with no wire. In the latter case, these holders (and the glass pipettes) must be completely filled with electrolyte fluid with no air bubbles to ensure good electrical coupling of the electrode to the amplifier. The back of the holder usually has a metal pin or wire coupling to the headstage. Suppliers: Siskiyou, A-M Systems, WPI.
- *Electrode storage*. Since piggyback multibarrels must usually be fabricated prior to experimentation, it is important to have a method of storing them in a way that not only prevents their accidental breakage but also prevents exposure to dust. A plastic slide box or Petri dish can be lined with modeling clay, which can be easily molded to support any type of electrode (e.g., single, multibarrel). However, care must be taken to prevent clay from getting inside capillary glass.
- *Iontophoresis*. Current injection can be used to eject charged molecules from glass electrodes or to create lesions with metal electrodes. Several commercial suppliers sell units designed for iontophoresis applications: NPI, Harvard Apparatus, Dagan.
- *Positional control*. The stable and precise positioning, placement, and movement of the recording electrodes are crucial to effective long-lasting recordings. In some

cases, the electrode holder plugs into the headstage directly, and the headstage itself is secured to an actuator or micromanipulator. When electrodes are connected to the headstage at a distance by silver/silver chloride wires, the headstage may be secured and immobilized, while the flexible wire connection allows the glass pipette to be moved independently. The glass pipette must still be secured to a holder of some kind, which can be attached to a micromanipulator. There are many types of micromanipulators available, which can incorporate or be used in conjunction with actuators. Motorized actuators have the advantage of being remotely driven if the subject is situated in a stimulus-controlled apparatus (e.g., sound isolation booth, or dark room for visual stimulus control). There are several types of micromanipulators that effect movement by different mechanisms (hydraulic pressure, stepper motor/screw-driven, piezoelectrical control). Several companies supply actuators: Siskiyou, Exfo Burleigh, Luigs & Neumann.

- *Adhesives*. Quickset cyanoacrylate glue (e.g., Loctite 404) and "cool-curing" dental cement (cures at room temperature; can be purchased from A-M Systems or a dental supply company) can be used to fabricate piggyback-style multibarrel electrodes and as general laboratory adhesives.

Recording Equipment

- *Amplifier*. There are many models and types of amplifiers available and costs vary widely depending on the features included. Most amplifiers require the use of a specific headstage, which may or not be included, but the specifications and dimensions of which should be considered prior to acquiring manipulator/actuator hardware.
- *Voltage window discriminator*. These can be used to detect signal event times by setting either an "above" or "within" window. Detection of an event generates a TTL pulse that can be sent to a processor and ultimately recorded as an "event time." The same function can be achieved with software analysis of the current waveform using Matlab, Igor, or Clampfit.
- *Processor*. If spike waveforms are to be recorded, the analog voltage signals must be digitized to be recorded by a computer. Several companies sell digitizers that may be designed to operate generically or specifically with other experimental apparatus. The software–hardware interface is a critical consideration in designing and setting up the apparatus.
- *Output monitoring*. The output of signals can be monitored in several different ways. One can couple the output of an amplifier to an audio speaker, which allows for auditory monitoring of both background noise and other signals. This can be immensely helpful in many types of recording, as the experimenter gets immediate feedback from the preparation that may not be obvious on a visual monitor. Amplifier output can be visually monitored with an oscilloscope (Tektronix or B&K Precision) or by viewing the digitized signal on a computer monitor using a software interface. While

some signals can be stored directly in an oscilloscope, limited sample capacity on oscilloscopes generally necessitates storage on a computer.

- *Faraday cage*. Electrical noise from power supplies and other sources can be dramatically reduced by enclosing the preparation in a "Faraday cage." A Faraday cage is simply a structure of metal or conductive screen that surrounds the preparation and is coupled to the ground of the amplifier. Cages are commercially available but can also be fabricated fairly easily and inexpensively in the lab.

Organismal Stabilization and Control

- *Stereotaxic apparatus*. This hardware serves the dual purpose of stabilizing the subject's head and body in a specific position and optimally allows for precise positioning of the electrode.
- *Antivibration table.* The stereotaxic apparatus itself can be secured to a floating vibration-isolating table (Technical Manufacturing Corporation). These tables are helpful for eliminating vibration at the preparation to enhance recording stability. Alternatively, very heavy platforms (typically made of stone or steel) can help eliminate some vibration.
- *Temperature control.* There are many reasons why it is important to monitor and maintain the body temperature of the organisms throughout the experiment. Most anesthesia treatments will disrupt temperature homeostasis. Changes in temperature dramatically affect neural responses, which can create a source of uncontrolled variability between experiments. For *in vivo* experiments, changes in temperature may also affect the comfort of the organism, which is always an important consideration. Many companies that specialize in physiological apparatus sell heating supplies that are controlled by a feedback circuit coupled to the subject's temperature monitor.

Microscopy

- *Dissecting microscopes*. Binocular dissecting microscopes are beneficial for any surgical or dissection procedures and the positioning of recording electrodes. Dissecting microscopes can be mounted on arms that allow them to swing in and out of place. These can attach to a standalone "boom" stand or can attach to a wall, or in some cases directly to an antivibration tabletop. It is important to consider the required working distance prior to purchasing expensive lenses. Objectives with larger working distances to allow for dissection and/or recording equipment are usually available.
- *Upright compound microscopes*. General-use compound microscopes (similar to what would be used in student laboratories) can used for viewing electrode tips and are optimal for fabricating piggyback multibarrel electrodes. Again, long-working-distance objectives are most convenient for this purpose.

Histology

There are different types of microtomes that can be used to create thin tissue sections. The thinness required varies on the intended purpose of the sections; for example, experiments requiring investigation of individual neurons (e.g., immunohistochemical studies) require sections that are close to the diameter of those cells (typically 10- to 40μm sections). For other purposes, such as investigating the location of electrode tracks or confirming recording sites, thicker sections may be used. Two types of microtome are typically used to create thin tissue sections: a cryostat or freezing microtome, where tissue is frozen before sectioning, or a vibrating microtome (vibratome) that mechanically vibrates a razor blade while sectioning tissue.

General Laboratory Equipment

The following is a list of general supplies and equipment that are helpful in performing the above experiments. For these items, there are many choices and options available, and this list is not comprehensive.

- Laboratory scales (sensitive to 1,000th of a gram)
- pH meter with calibration buffers
- Adjustable-volume micropipetters (10, 20, 200, and 1,000 μL) and tips
- Microcentrifuge
- Eppendorf tubes
- Conical centrifuge tubes
- Tube racks
- Weigh boats and weigh paper
- Scintillation vials
- Glass bottles and caps
- Graduated cylinders
- Beakers
- Source of purified water, such as a reverse osmosis unit (e.g., from Millipore)
- Hot plate/magnetic stir plate
- Magnetic stir bars
- Sylgard (Dow Corning), a liquid silicone polymer that cures to a gel-like solid that can used to fill dishes, chambers, and so forth. Sylgard is often used to pin down organisms or tissue for dissection and/or recording.

Reagents and Chemical Supplies

- Physiological salts (Fisher Scientific and SIGMA are typical sources): NaCl, KCl, $CaCl_2$, $MgCl_2$, K acetate, monobasic and dibasic phosphate salts for making phosphate buffers (10× PBS pellets or liquid stocks are also available).
- Other buffers: Tris or HEPES

- HCl, NaOH, KOH (used for adjusting pH)
- Glucose (or other sugars; type depends on organismal requirements)
- Paraformaldehyde, a commonly used aldehyde for fixing tissue to be used for immunohistochemistry or other histological processing. Available as a solid (prills), powder, or liquid.
- Biotin/Biocytin/Neurobiotin and ABC Staining Kit (Vector Laboratories), one example of an anatomical tracer that can be included in glass recording pipettes to mark recording sites.

Anesthetics

- Ketamine (Ketaset, Fort Dodge Animal Health): requires DEA license
- Sodium pentobarbital (SIGMA, others): requires DEA license
- Urethane (SIGMA)
- Inhalant anesthetics (e.g., isoflurane or halothane) are also available from several companies.

Frequently Asked Questions

Q. Why do I have trouble finding my target area from experiment to experiment?

A. This is a common trouble spot for *in vivo* recording experiments, particularly when the target area is not superficial enough to be easily visible. The key to consistency between experiments is to minimize variability as much as possible. This may include using organisms of similar age or size, keeping the organismal position (i.e., the head angles) as similar as possible, and making sure that all related components of the apparatus are in the same position from preparation to preparation, including the floating table, the platform that holds the organism, the organism itself, the electrode holder, and the electrode. Using anatomical or other reference marks on the skull that are in a consistent location as a starting point to orient electrode penetrations can also be very helpful. In practice, diligent use and evaluation of histological markers over several experiments can enhance targeting accuracy.

Q. Why I am having trouble getting consistent electrode resistances?

A. Several problems can arise when using glass pipettes as electrodes. The most common problem is variability of electrical resistance. Electrode resistance is a key feature that determines one's ability to monitor "background" activity or isolate single neurons, or both. If the resistance is too high, it can be difficult to record any signals at all. Conversely, individual neurons cannot be well isolated if the resistance is too low. Sudden changes in electrode resistance within a single penetration can inform the experimenter of various electrode conditions. For example, if resistance suddenly increases, this indicates that the electrode may be blocked; this blockage may be reversed by injecting current. A sudden drop in resistance may indicate that the electrode tip has broken. This can be particularly frustrating because the cause of breakage may not be immediately obvious. A thin piece of skull or meninges that was not removed during surgery is a common cause of electrode tip damage. Very careful removal of these materials helps avoid this problem. Electrodes must be maintained within the optimal range (typically 1–10 MΩ for glass) to collect data as efficiently as possible. Frequent electrode replacement is common with glass pipette recording.

Another cause of variability with glass pipettes arises from the use of pullers that have a metal heating filament. The variability can appear as a change in shape or electrical resistance (or both) over time, even though the same heating and pull settings have been used as for previous pipettes. This may occur because the metal of the heating filament may become oxidized over time, which changes its heating properties. Thus, it is important to frequently check the resistance and shape of electrode tips, and the heating and/or pull settings may need to be adjusted over time to compensate for changes to the filament. After prolonged use, the heating filaments may need to be replaced.

Q. How can I reduce or eliminate sources of interference or "noise" from my recordings?

A. This is a common (and sometimes extremely frustrating) problem with any type of electrical recordings. Common sources of noise include cellphones, fluorescent lights (common in laboratories), cooling fans, and electrical "line noise" from the power supplying the laboratory and its equipment. Some of these can be as simple to fix as just removing the source of noise (i.e., turn off the lights in the soundproof booth while recording). Other types of noise are not so easy to fix or identify. In the case of 50- to 60 Hz noise coming from an AC outlet, electronic filters (e.g., notch filters) can be used with various settings to eliminate this frequency of noise. Some amplifiers have built-in filters. External noise reduction circuits can be coupled to the amplifier output prior to monitoring on an audio monitor, computer, or oscilloscope. Noise can be exacerbated by bad electrical connections and/or a faulty wire. In some cases, the only way to troubleshoot noise is to disconnect and replace wires one at a time with a tested functional wire. It is also important to maintain silver/ silver chloride wires, which may need to be periodically rechlorided or replaced. It is a good idea to keep a stock of replacement wires on hand so that they can be quickly replaced in case they suddenly break or become noisy during an experiment. Another common source of noise is grounding issues (e.g., the presence of ground loops, incorrectly placed ground wires, faulty wires). It is crucial to make sure that everything that needs to be connected to ground actually is. It is also good practice to have all ground wires connecting to a common point, which may be as simple as a screw or metal block near the preparation that is itself securely connected to the amplifier ground. Finally, it is important that there is always a complete recording circuit: Be sure that the ground wire in the organism (or recording bath) stays properly connected.

Q. Why am I having trouble "holding" neurons while recording?

A. A common problem that occurs during *in vivo* single-unit extracellular recordings is that neurons are "lost" before experiments are complete. While some neurons will be lost seemingly at random (or for no cause at all), there are several techniques to help keep neurons isolated for longer periods of time. First and foremost is to stabilize the recording conditions as much as possible. Some helpful tips include using a vibration-isolating table and physically securing the organism with a head post or other rigid restraint device.

Other important considerations include exactly how neurons are isolated. For example, one should advance very slowly through the target tissue area and frequently "clean" the electrode by passing current through it between recording from isolated neurons. Neurons that are isolated in haste or by accident are more likely to be lost rapidly than neurons that were isolated during a slow approach. Another helpful suggestion is to carefully watch the background noise and allow time for the tissue to "settle" while advancing the electrode (advance slowly, and make frequent pauses), as neurons may be lost during this settling period. It may also be advantageous to gently reverse the electrode 5–20 μm once a high signal-to-noise ratio is achieved. In other cases, if isolation signals start to become smaller, it may be necessary to slowly advance the electrode to maintain isolation. As with all things, practice makes perfect: Consistency and efficiency will increase if you persist and keep trying.

Further Reading and References

The axon guide: A guide to electrophysiology & biophysics laboratory techniques (3rd ed.) (2008). Sunnyvale, CA: Molecular Devices/MDS Analytical Technologies.

Coleman, W. L., Fischl, M. J., Weimann, S. R., & Burger, R. M. (2011). GABAergic and glycinergic inhibition modulate monaural auditory response properties in the avian superior olivary nucleus. *J Neurophysiol* 105: 2405–2420.

Havey, D. C., & Caspary, D. M. (1980). A simple technique for constructing "piggy-back" multibarrel microelectrodes. *Electroencephalogr Clin Neurophysiol* 48: 249–251.

Köppl, C., & Carr, C. E. (2008). Maps of interaural time difference in the chicken's brainstem nucleus laminaris. *Biol Cybern* 98: 541–559.

Tabor, K. M., Coleman, W. L., Rubel, E. W., & Burger, R. M. (2012). Tonotopic organization of the superior olivary nucleus in the chicken auditory brainstem. *J Comp Neurol* 520: 1493–1508.

Multi-electrode Recording of Neural Activity in Awake Behaving Animals

Samantha R. Summerson and Caleb Kemere

Introduction

Perhaps one of the most important questions we seek to understand when studying the brain is how neural circuits process information. One of the key ways of addressing this question is to record the activity of individual neurons during well-controlled behaviors and/or presented stimuli. By measuring the spiking activity of the neurons during sensation or action, we can build models of the output of a given neural circuit. Then, by building a hierarchy of models, we can understand the broader flow of information.

Traditionally, experimentalists have acquired data and built these models based on data obtained one neuron at a time. In these types of experiments, activity of multiple neurons is assembled over many sessions during which stimulus parameters and/or animal behavior is carefully monitored and/or controlled. Isolating the signal of a single neuron requires precisely positioning the tip of electrode in the proximity of the soma, as described in Chapter 3. Then, a successful experiment requires maintaining the electrode stably in that position long enough to acquire whatever data are needed. For some brain regions this can be a painstaking process in which a large fraction of an experimental session is spent locating the neurons to be studied. In other cases, the challenge is that the action potentials of multiple neurons are detected by the electrode, requiring ultraprecise adjustment to isolate the activity of one from the rest. The impedance of the recording electrode, which is largely determined by the surface area of the recording region (Cogan 2008), predominantly determines where a given experiment lies along the continuum from zero-or-one neuron (high-impedance electrode) to one-neuron-from-many (lower impedance).

Simultaneous recording of neural activity using multiple electrodes has the obvious potential advantage of reducing the amount of time required to acquire data. In practice, the challenges of precise positioning mean that one cannot simply scale electrode count by sequential placement of individual impermanent electrodes. By the time the nth electrode is targeted to a neuron, the signal on the first electrode has likely been degraded. Instead, many systems neuroscience experiments employ chronically implanted multichannel electrode arrays in which permanent electrodes are placed concurrently, in parallel. The multi-electrode approach enables the activity of many neurons to be recorded simultaneously and chronic implantation can provide stable signals across multiple days (Chestek et al. 2011, Fraser and Schwartz 2012, Kentros et al. 1998, Tolias et al. 2007). As we will discuss later, chronic implantation typically entails the use of lower-impedance electrodes, requiring signal processing-based isolation of individual neurons by the shape of their action potentials ("spike sorting"). However, when electrode sites are close enough that the signals of individual neurons are detectable on multiple channels, confidence in the quality of unit isolation can actually be higher than in all but the highest-quality single-channel electrode recordings (Ecker et al. 2010, Gray et al. 1995, Harris et al. 2000, Henze et al. 2000).

In addition to enabling faster "single-electrode" science, simultaneously recording the activity of multiple neurons enables the study of important aspects of neural circuits that cannot be observed from a single neuron at a time. To understand this, we must consider how the activity of an individual neuron varies. When repeated stimuli are presented, the exact pattern of action potentials generated by a neuron is variable. In the simplest model, we might describe this variation as arising from two sources, "signal" and "noise." The "signal" component of variation reflects how changes in the stimulus or the animal's action modulates brain activity (e.g., a visual receptive field or a preferred direction of motion). The "noise" component reflects variability that arises from the randomness in synaptic transmission, membrane fluctuations, and so forth (Averbeck et al. 2006, Harris and Thiele 2011). Thus, to distinguish signal from noise, we need exquisitely precise control of experimental and behavioral parameters, as well as repeated trials. However, even when these conditions are met, variations in the internal state of the brain (e.g., attention, alertness, or motivation) can modulate neural activity in uncontrolled ways (Averbeck and Lee 2004, Chalk et al. 2010, Goard and Dan 2009, Harris and Mrsic-Flogel 2013; Harris and Thiele 2011, Harris 2005, Karlsson et al. 2012, Lee and Dan 2012a, Niell and Stryker 2010, Ponce-Alvarez et al. 2012). When multiple neurons are recorded simultaneously, it begins to be possible to use statistical models to disambiguate brainwide variation from the fluctuations of individual neurons (Briggman et al. 2006, Chen et al. 2011, Churchland et al. 2007, Kass et al. 2005). In addition, by recording the activity of multiple neurons in an anatomically connected circuit, we can begin to understand exactly how neural ensembles transform patterns of information. Thus, multi-electrode recording has emerged as a key tool in expanding our understanding both of how neurons represent the world and how they generate these representations (Buzsáki 2004, Stevenson and Kording 2011).

Multi-electrode Recording Arrays

Several multi-electrode recording technologies are employed for chronic neural recording. We can divide them into two broad categories: commercially manufactured electrode arrays and customized "microdrives." To form a complete system, these multi-electrode devices are coupled with multichannel neural recording systems.

Commercially manufactured rigid electrode arrays come in three general categories: microwire arrays (Nicolelis et al. 2003), micromachined electrode arrays (Normann et al. 1991), and multisite silicon probes (Drake et al. 1988, Normann et al. 1991, Vetter et al. 2004). As shown in Figure 4.1, microwire arrays (Microprobes for Life Sciences, Inc, Gaithersburg, MD; Tucker-Davis Technologies, Alachua, FL) and micromachined electrode arrays (Blackrock Microsystems, Salt Lake City, UT) are typically structured as a "bed of nails," where long, narrow electrodes are insulated except for the very tips. Were these electrodes to be rigidly attached to the skull, micromovements of the brain (e.g., arising when animals vigorously interact with their environments) would cause rapid degradation in the quality of recorded signals. Consequently, for stable chronic recordings, a "floating" architecture is often used, in which a flexible tether connects the implanted electrode array to the connector used to record signals (see Fig. 4.1c). Typical

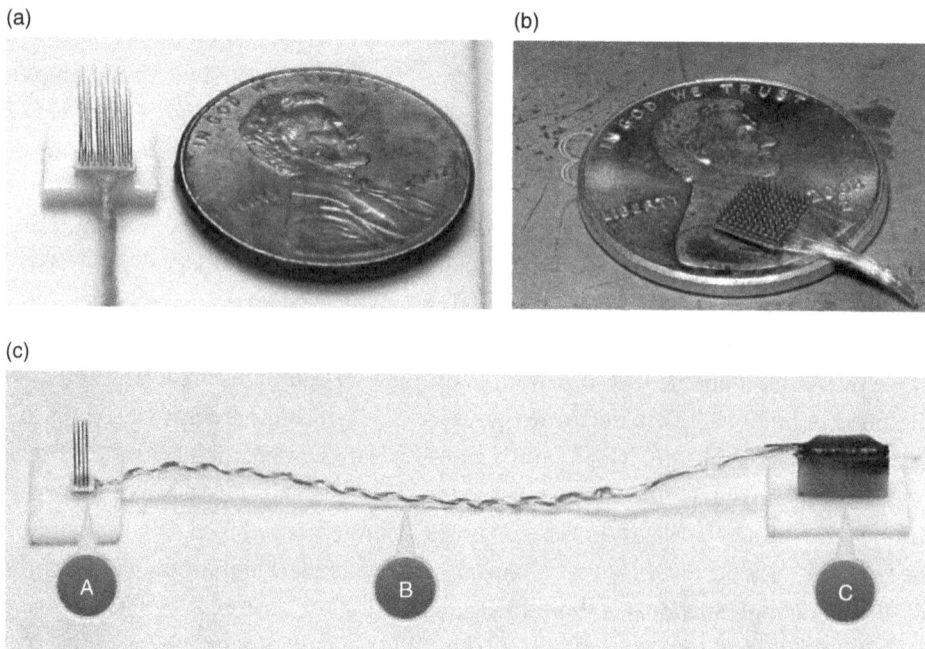

FIGURE 4.1 Two variants of a chronically-deployed electrode array, both structured as a "bed of nails." a: Floating microwire array (Microprobes for Life Sciences, Inc.). b: Silicon micromachined electrode arrays (Blackrock Microsystems, Inc., photo by Bryan William Jones, used with permission). c: A flexible tether (B) connects the implanted electrode array (A) to the connector used to record signals (C). Photos (a) and (c) by Martin Bak, MicroProbes for Life Science, Inc., used with permission.

interelectrode distances are 100–500 μm, meaning that signals from individual neurons are present on at most one electrode. By specifying the depth of the electrodes, these topologies can be customized for recordings targeting a particular layer of neocortex (these arrays are typically not useful for subcortical structures).

In contrast to targeting a particular plane of neurons, silicon probes have multiple recording sites along the depth axis of the probe (Fig. 4.2). This architecture has enabled local field potential recordings along the dendritic arbors of neurons, which are critical to understanding the architecture of vertically organized structures in the hippocampus and neocortex (Bragin et al. 1995, Buzsáki et al. 1992, Ylinen et al. 1995). Newer electrode topologies that place recording sites in an array at the tip of the electrodes have proved more successful in chronically recording the activity of single units (Mizuseki et al. 2009, 2012, 2014)—the key innovation that enabled this success is the ability to move the electrode array (Vandecasteele et al. 2012).

In contrast, in most chronic recording experiments rigid electrode arrays are initially implanted and then not moved. Unfortunately, precise targeting of recording sites to individual neurons is impossible during implantation. Thus, the most common approach for obtaining high-quality multi-electrode recording of many single neurons

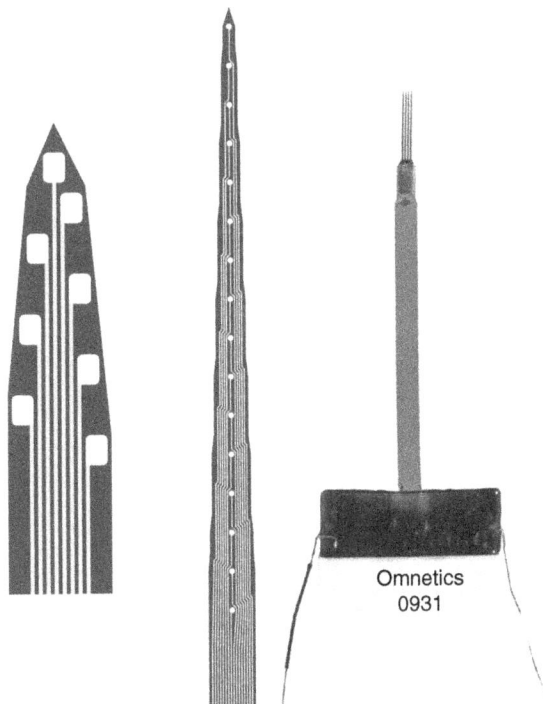

FIGURE 4.2 Silicon probes have multiple recording sites along the depth axis of the probe. Right: Actual device. Left and center: Sample recording site layouts. In general, the electrode pattern chosen depends on the neurons being targeted and the experimental paradigm. Drawings and image by Neuronexus, Inc., used with permission.

uses microdrives (Fig. 4.3) (Gray et al. 1995, Jog et al. 2002, McNaughton et al. 1983, Wilson & McNaughton 1993). While there is limited commercial manufacturing of microdrives (Neuralynx, Inc, Bozeman, MT), including recently introduced drives that allow the depth of rigid electrode arrays to be adjusted after implantation (NeuroNexus, Inc.), most are still manufactured within individual laboratories. Microdrives are typically used with two- or four-channel electrodes ("stereotrodes" or "tetrodes"). Remarkably, the procedure for manufacturing these electrodes—twisting nichrome or platinum–iridium wires together, softening the insulation so that they will adhere, and cutting and electroplating the ends—largely follows the same procedure developed more than 30 years ago (McNaughton et al. 1983). A typical microdrive will have between 4 and 40 tetrodes, with smaller microdrives used for recordings in mice (where there is an implant weight limit of ~5 g) (Voigts et al. 2013) and larger microdrives used for recordings in rats (limited to ~40 g) (Kloosterman et al. 2009) or even monkeys (Tolias et al. 2007). The canonical design presented here for a 16-tetrode microdrive has a mass of ~35 g and can be expected to record the activity of >100 neurons in dorsal hippocampus (see the end of the chapter for a detailed, step-by-step protocol to manufacture a 16-tetrode microdrive).

Surgical implantation of chronic recording electrode arrays in rodents is beyond the scope of this chapter (see Vandecasteele et al. 2012). However, several factors are discussed below that merit special attention: (1) the ground and/or reference screw, (2) minimizing damage to cortex to minimize inflammation, and (3) ensuring good adhesion to the skull for stable recording.

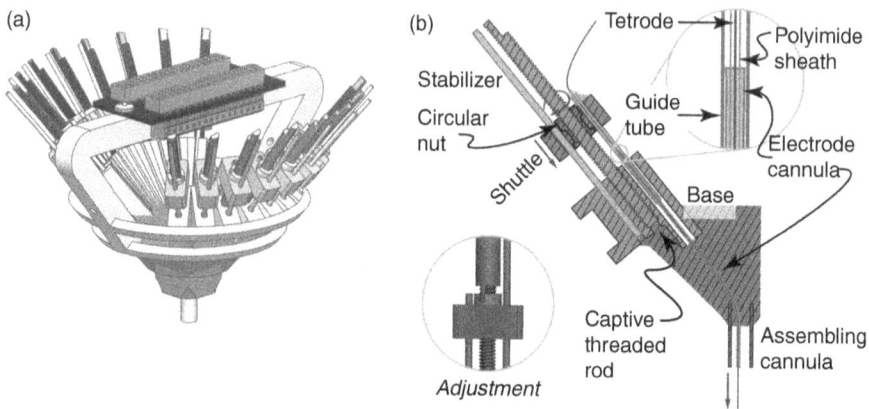

FIGURE 4.3 Microdrive arrays are typically hand-assembled. By using a very small threaded rod for adjustment of each individual electrode, precise positioning can be achieved. a: Sketch of completed device showing 16 independently movable microdrives intended for tetrodes. b: Drawing of one microdrive assembly. In this design, shuttles are made by hand and coupled with 0–80 threaded rod, a custom 3D-printed base, and individual stainless-steel cannulae to form a complete drive.

Multi-electrode Data Acquisition Systems

A block diagram for a traditional multi-electrode recording system is shown in Figure 4.4. Neural activity recorded from each electrode is amplified by a small circuit board plugged into the electrode connector (the "headstage"), transmitted along a tether, further amplified, digitized, and then processed by custom software on a personal computer. In the most recently available neural data acquisition systems (e.g., Neuralynx, Inc.; Tucker-Davis Technologies, Inc.; Plexon, Inc.; Blackrock Microsystems, Inc.; Ripple, LLC), the complete amplification and digitization process occurs in the headstage (Harrison 2008), reducing the number of wires in the tether from one per channel to a dozen or less.

Practice of Multi-electrode Recording

Multichannel Neural Activity

Software designed for recording multichannel neural activity will typically include the ability to monitor both local field potentials (LFPs) and spiking activity. The LFP is the lower-frequency (typically less than ~400 Hz) component of the acquired signal (see also Chapter 7). The low-pass filtering removes much of the signal energy that arises from the action potentials of nearby neurons and leaves a signal thought to be primarily driven by synaptic currents. Depth-dependent changes in LFP patterns are often extremely useful for fine positioning of recording electrodes. LFP is typically visualized as a series of scrolling records, as shown in Figure 4.5.

Action potentials are typically detected using an amplitude threshold. The recording software enables the threshold level to be optimized for each channel of the system. Neural "snippets"—windows of electrode signal around the times of threshold crossing (typically ~0.25 ms before and ~1 ms following)—are displayed for each electrode channel. For tetrode recordings (and other multichannel electrodes where more than one channel is associated with a micro-scale brain location), action potentials from a single neuron can often be detected on multiple channels. Thus in these cases, a threshold crossing on one channel triggers snippet acquisition for all associated channels. Furthermore, for tetrode recordings, the snippet display typically also includes a "cloud" view where the peak values on each channel are plotted against each other. This perspective allows for rapid estimation of the quality of signal isolation for individual neurons, which appear as isolated clouds. Both snippet and cloud views display not only the most recent threshold crossing detected but also some history (see Fig. 4.5). Proper selection of the threshold is different during recording and during adjustment and is discussed below.

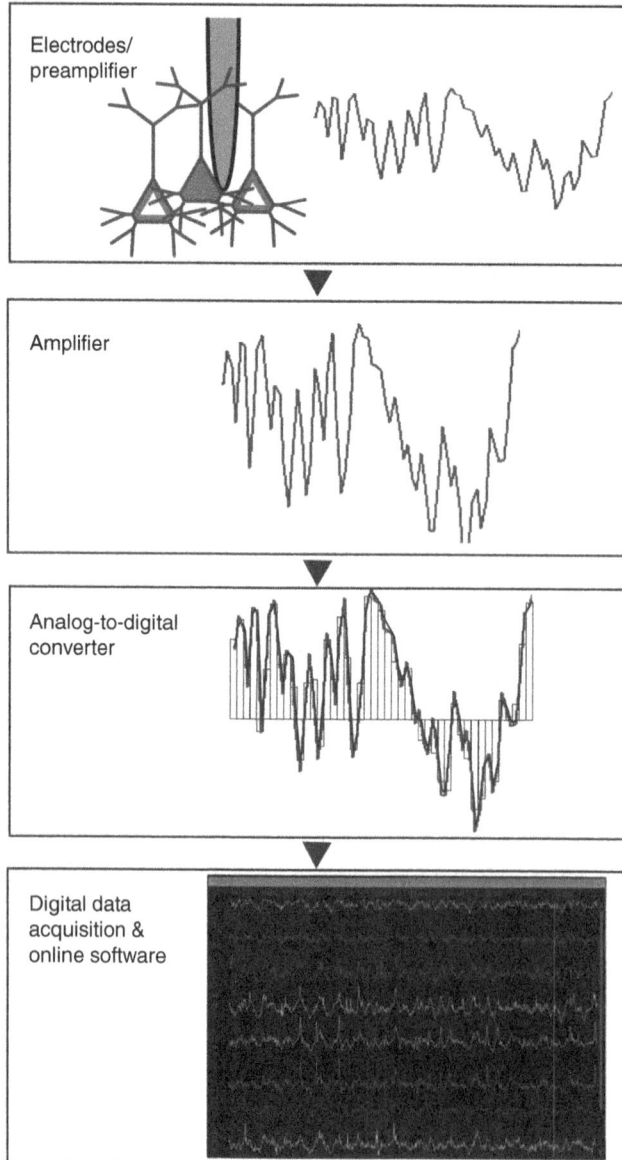

FIGURE 4.4 In a multi-electrode recording system, data acquisition comprises several steps. Top: The signal is acquired from implanted electrodes, (middle top) amplified from microvolt levels to a range appropriate for analog-to-digital conversion, (middle bottom) digitized, and (bottom) stored and displayed on one or more computers. In most traditional multi-electrode recording systems, only the first step occurs at the animal's head, where a preamplifier drives the signals up a long multiwire tether to the amplifiers. Custom integrated circuits have enabled an emerging trend in which the amplification and digitization take place on the animal's head, resulting in a significant reduction in the number of wires in the tether.

FIGURE 4.5 A screen shot of the Open Ephys GUI shows two open panels. Left: LFPs are visualized as a series of scrolling records, with a different color for each tetrode. Right: Waveform snippets acquired from each tetrode are plotted, and the peak amplitudes of each snippet are juxtaposed pairwise to form six two-dimensional clouds of points. By visualizing the pairwise signals, an experimenter can quickly assess whether action potentials of one or more neurons can be isolated. (Courtesy of Dan Johnston and Joshua Siegle, used with permission.)

Ground and Reference

During neural recording experiments, the configuration of the data acquisition system can have significant effects. All voltage measurements are differential, requiring a return path for the signal, and neural signals are no exception. The electrode against which the voltage of the recording electrode is measured is called the return electrode. In many cases, the return electrode is also connected to the ground for the neural recording electronics. A good return electrode is essential for high signal quality. A common choice is a conductive, biocompatible (stainless steel or titanium) screw implanted over the cerebellum. Care should be taken to ensure not only proper placement, with the bottom of the screw in contact with cerebrospinal fluid, but also a quality conductive solder joint between the screw and the connecting wire to the headstage (Fig. 4.6). For stainless-steel screws, an appropriate flux solution (i.e., N-3, LA-CO Industries, Inc., Elk Grove Village, IL) is required for the solder bond to form. The postimplantation quality can be tested by measuring the DC resistance between the animal's ear pinched in a saline-soaked gauze pad and the ground/reference connection on the headstage. Values greater than ~20 KΩ suggest that some portion of the return path is defective.

After the signal on each channel is amplified and digitized, a second type of referencing can be applied. This is simply accomplished by differencing the signals between a signal electrode and one designated as a reference. Reference electrodes are typically chosen based on their positioning far from neural sources—fiber tracts such as the corpus callosum are common choices. This digital referencing can enable action potential recording despite persistent external sources of noise that are common to multiple electrodes. As shown in Figure 4.7, large-amplitude noise will saturate the analog-to-digital conversion, resulting in maximal digital values, and thus digital referencing will not

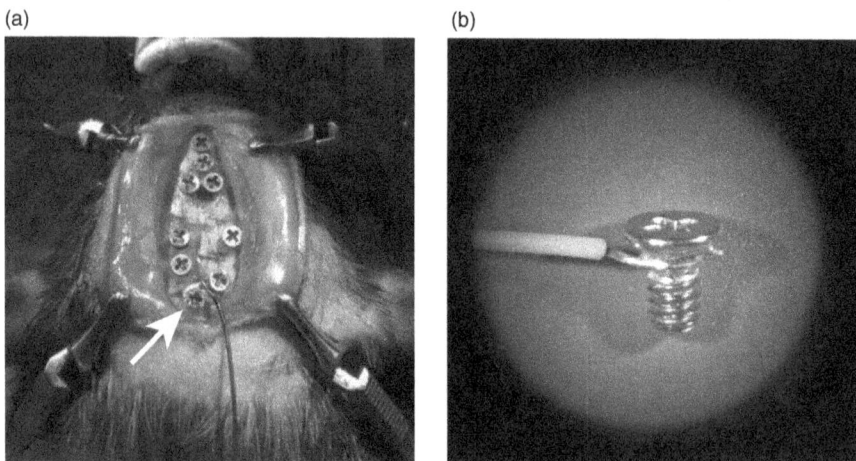

(a)   (b)

FIGURE 4.6 a: During surgery, a stainless-steel ground screw (*arrow*) is implanted into the skull over the cerebellum at the same time as implant fixation screws. b: A simple way to create a conductive connection to stainless steel is to wrap a stainless wire around the top threads of a fixation screw, and then apply solder using an appropriate flux.

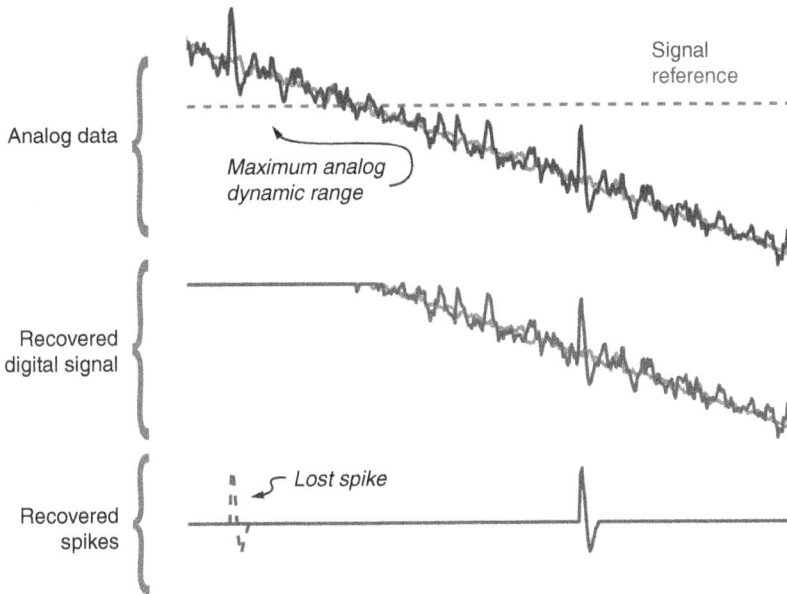

FIGURE 4.7 Digital referencing of the signal from neural data electrodes to another electrode(s) in the brain can sometimes mitigate the effects of large noise sources. In this example, a large noise event has perturbed both the signal and reference electrode channels in the analog data. Even though the level of nonsignal noise is much higher than the amplitude of the action potentials, the second spike can be recovered after subtracting the local reference. However, in the case of the first spike, both signal levels have exceeded the maximum input to the ADC, and thus small variations are erased.

be able to recover small signals. LFP signals are often present on electrodes used for digital referencing. Thus, care must be taken when recording/analyzing LFPs that might include signals from both recording and reference electrodes.

System Setup

Before plugging in the animal for the first time:

- Visually check the tether and headstage. Examine headstage(s) and cable for damage. Ensure that long tethers are mechanically supported (typically with elastic or counterweight systems) in such a way that animals will be able to move freely within the experimental environment. Many investigators additionally use a commutator to allow for animal rotation.
- Use a multimeter to measure the DC voltage between the pins of each headstage that will connect to electrodes and the pin that will connect to the ground screw. A non-zero voltage reflects a hazardous condition and is commonly caused by incorrect cable connections or damage to the headstage(s).
- Verify signal acquisition. Turn on data acquisition software. Plug a signal generator into the headstage(s). Verify that there is minimal noise and that 10 to 100 µV, 40 to 1,000-Hz sinusoidal signals can be properly detected. Most neural data acquisition

systems enable real-time audio monitoring of received neural data. This is an invaluable tool for both setup and recording. Under most circumstances, environmental sources of noise can be detected at this stage of setup.

- Eliminate noise. Common sources of noise are "ground loops" and nearfield radiation. We commonly think of neural data being transmitted along one wire, but just as in our discussion of referencing, all signal paths must be a circuit. Ground loops arise when multiple high-impedance return paths for signals are shared with other equipment. In this case, it is very common to detect 50/60-Hz noise coupled from power supplies into the neural data. Ground loops can be alleviated by (1) ensuring there is only one conductive ground path from the animal to the recording system (avoiding, e.g., a grounded, conductive behavioral environment) and (2) using only DC-powered equipment and ensuring that DC power is "low ripple." Radiative sources of noise are more insidious than ground loops. These arise because neural recording electrodes act like antennas. Thus, local sources of radiative energy can induce noise. Note that this noise need not be within the frequency band of neural activity (i.e., <20–30 kHz); enough power will overcome the anti-aliasing filters and be modulated into the neural activity band. Common sources of radiative noise include switching power supplies (commonly found in DC power adaptors), fluorescent lights, and other scientific equipment. These noise sources can be alleviated by electrically shielding either the noise sources or the neural recording environment with grounded conductive material such as copper mesh or aluminum foil.
- Verify experiment. Check for proper functioning of any behavioral control or measurement apparatus (e.g., operant conditioning, video monitoring). Verify that these data are properly synchronized with neural data.

Electrode Positioning

Disruption or removal of the membranes that cover the brain surface (thick dura and thinner dura and arachnoid) inevitably result in an inflammatory response (Polikov et al. 2005). One of the consequences of this inflammation is expansion of the cortical sheet that decreases in the days (up to two weeks) following surgery. Thus, for electrodes lowered during or immediately following implantation, the depth of recording sites relative to target neurons will change over time. If this movement causes the electrodes to move through and beyond the target recording depth (i.e., as reduction of inflammation shrinks the tissue above), resulting damage can drastically reduce the quality of final recordings (Fig. 4.8a). Consequently, the final positioning of recording electrodes to target neurons should be postponed until after inflammation has subsided.

Electrode movement can also be irritating to tissue, causing further inflammation and even seizures in severe cases. Thus, care should be taken to move electrodes slowly. Specifically, during daily adjustments, the electrode should be lowered or raised slowly, in a smooth manner, at a rate of ~10 µm/s. Even when electrodes are moved slowly and carefully, some tissue displacement will take place (see Fig. 4.8b), resulting in drift of the final location overnight of up to 25% additional displacement. This can be ameliorated by retracting the electrode a small amount after advancing it, but care should always be taken that this overnight drift will not move the electrode beyond the desired location. To reduce the effects of inflammation and tissue displacement, the distances electrodes are advanced should follow an exponentially decreasing pattern.

When advancing electrodes, care should be taken to securely restrain the movement of the implant without adversely restraining the animal. The implant should be held close to the base to prevent abrupt movements from applying a torque that could cause deimplantation. During implantations, the use of special adhesives and bone-preparation agents such as Metabond (Parkell, Inc.) applied between skull and methylmethacrylate implant material can reduce these risks. The adjustment tool should also be held carefully to prevent sudden relative movements that could damage electrodes. It is easiest to make depth adjustments in sleeping animals. However, a trained investigator will hold the implant firmly with one hand and the adjustment tool with the other, while still allowing the animal to move slowly and securely. When animals are being adjusted, it is helpful that they be contained in a low-static environment. Either a raised platform or a four-sided box with high walls and no top is an appropriate choice (exploring animals will typically try to climb out of low-walled environments). Allowing the

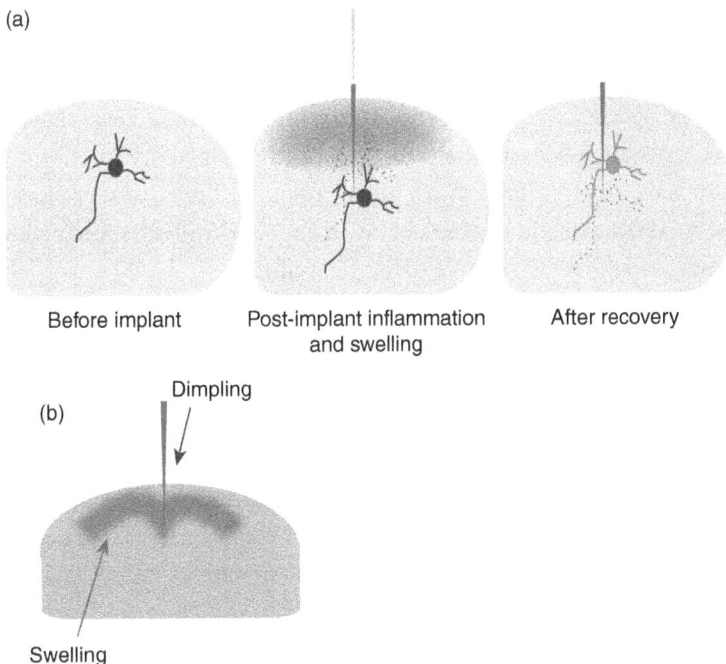

(a)

Before implant   Post-implant inflammation   After recovery
and swelling

(b)

Dimpling

Swelling

FIGURE 4.8 a: Implantation can cause swelling, which shifts the relative position of target neurons. If the electrode is lowered to the target too early, deswelling can result in the electrode passing through and damaging or destroying the target area. b: Even when electrodes are moved slowly and carefully, some tissue displacement in the direction of movement and potentially slight damage will take place. Thus, electrodes should always be moved slowly and gradually.

animal to explore the adjustment environment thoroughly before surgery (or prior to electrode adjustment) can reduce novelty-induced movements.

The process of electrode advancement and positioning requires experience. Thus, for experiments with very specific targets, new investigators should carefully note various patterns of neural firing and LFP and depths during advancement, comparing measured depths with an atlas. After final location is verified in postexperiment histology, these notes will facilitate subsequent accurate positioning. While patterns of neural activity are quite depth-dependent, the investigator must keep careful track of the depth of each electrode, ideally with at least 50-μm precision.

Electrode positioning in the days following surgery typically takes the following pattern:

Day 0: Advance electrodes into brain. Use audio monitor to listen to signal from each electrode during movement. Pause periodically to assess LFP and audio signals. If possible, record depth at which the brain surface is reached (often a subtle "pop"). Continue to advance until first action potentials are detected.

Day 1: Advance reference electrode to a quiet location (e.g., white matter). Assess signal on all other electrodes, then advance ~100–200 μm to find large action potentials. Note the quality of each channel of each electrode for a baseline against subsequent changes (which might reflect damage).

Days 2–5: Advance all electrodes to ~500–1,000 μm above final target depths. Move no more than 1–2 mm per session, and move furthest on earlier days. Observe LFP and spiking phenomena specific to different lamina (e.g., thalamocortical spindles in deeper cortex, decreased background activity in corpus callosum, sharp wave ripple complexes in hippocampus). These can serve as a check for measured depths, and conflicts between expected and observed depths of these features can reflect the degree of tissue swelling.

Days 6–9: Observe LFP and spiking on each electrode to assess the stability of the recording depth. Begin moving toward target area. Keep in mind that because of tissue stretching, electrodes will travel an additional 10–20% overnight. One strategy to minimize this uncertainty is to advance and then partially retract. The goal for final adjustment is to move ~50–75% of the remaining distance each day; as movements of less than ~25 μm are impossible, this will place the electrode at the target in 3–5 days. With tetrode recordings, extracellular action potential waveforms with peaks of 200–400 μm are expected, and occasionally values >1 mV can be stably recorded.

Days 10+: Observe LFP and spiking on each electrode. Stable implants should not exhibit laminar differences in patterns—changes in depth may reflect skull changes that will result in deimplantation. Recording quality is typically stable for 5–10 days but eventually will begin to decrease, reflecting local foreign-body responses. Advance electrodes as desired to find new neurons.

The following is an example case of the process for lowering electrodes to the pyramidal cell layer of the dorsal aspect of hippocampal area CA1. Note that cortical swelling typically depresses the dorsal hippocampus ~300–400 μm. Thus, we expect target depths initially to be deeper and then to become shallower. Depending on the degree of inflammation, this can take 5–12 days.

> *Day 0:* Advance electrodes into brain. Continue to advance until first action potentials are detected. Depending on exact coordinates, these neurons will often have motor, somatosensory, or visual receptive fields that can be activated by movement, stroking the animal, or flashing a light.
>
> *Day 1:* Advance reference electrode to corpus callosum (expected at a depth of 2–2.5 mm). Assess signal on all other electrodes, then advance to top of corpus callosum (expected at similar depths).
>
> *Day 2:* Advance all electrodes beyond bottom of corpus callosum. Very quiet background activity will begin to change as hippocampal sharp wave ripples (present during sleep and immobility) and ensemble activity become more apparent. If single excitatory neurons or dying cells are detected, stop and quickly retract electrode 100–200 μm. Move electrodes forward until ripples are clearly audible, then retract 300 μm. Figure 4.9 depicts signals expected at different hippocampal laminae.
>
> *Day 3–6:* Observe LFP and spiking on each electrode to assess swelling and stability. Again move electrodes forward until ripples are clearly audible, then retract 300 μm. After a week, or once deswelling (stronger sharp wave ripples than expected) is observed, begin final adjustment.
>
> *Day 7:* Advance electrodes until ripples are apparent and 40- to 60-μV peak multi-unit activity is present. Pull back 75 μm. Note that most ripples will be accompanied by downward-going sharp waves above stratum pyramidale.
>
> *Day 8:* Advance electrodes until multi-unit activity grows and single units can begin to be distinguished (100- to 200-μV peaks). Pull back 25–50 μm.
>
> *Day 9:* Advance electrodes 25 μm. Note that for this small of an adjustment, sometimes no immediate effect is noticeable.
>
> *Day 10:* Advance electrodes in 25-μm steps as required. When electrodes are within stratum pyramidale while ripples are prominent, sharp waves are not present.

Daily Practice of Neural Recording Experiments

In most awake behaving rodent experiments, the first steps of a typical experimental day will involve retrieving the animal from the vivarium, placing it into the container used for electrode positioning, plugging the animal in, and checking and setting parameters for the neural recording software. Ideally, the software should be set to

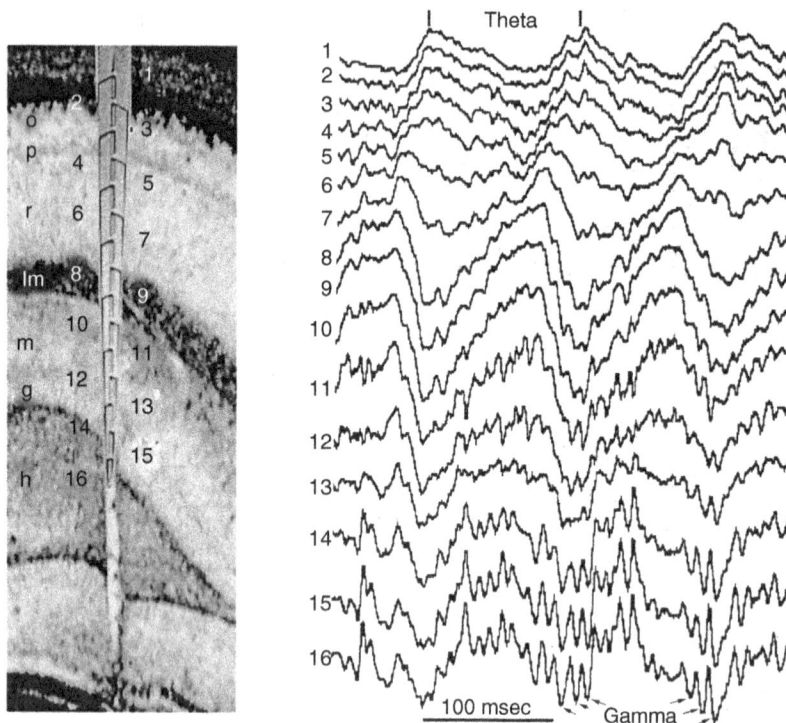

FIGURE 4.9 LFP patterns expected at different hippocampal laminae. (Reproduced from (György Buzsáki, 2002) with permission.)

record broadband data from all electrodes. This enables all LFPs and spikes to be extracted after the experiment. However, the data requirements (e.g., 14 GB/h for 64 channels sampled at 30 kHz) can sometimes be excessive. Hence sometimes investigators simply choose to save LFP—filtered and decimated to, for instance, 1,500 Hz—and thresholded snippets. If this is the case, care must be taken to select the proper thresholds for spike detection. As shown in Figure 4.10, if thresholds are set too high, the activity of potentially isolatable single neurons will be lost. If thresholds are set too low, the refractory nature of spike sampling can result in only the very beginning of large action potentials being detected. For most recordings, a safe range for threshold values is 45–75 μV.

The experimenter should follow the following steps:

1. Check integrity of headstages, tethers, and so forth. If anything has been reassembled, use a multimeter to measure voltage between recording channels and ground (a non-zero voltage is a hazardous sign of a misconnection).
2. Check that behavioral apparatus and behavioral recording system (e.g., video) are functional.
3. Observe animal for signs of pain or distress.

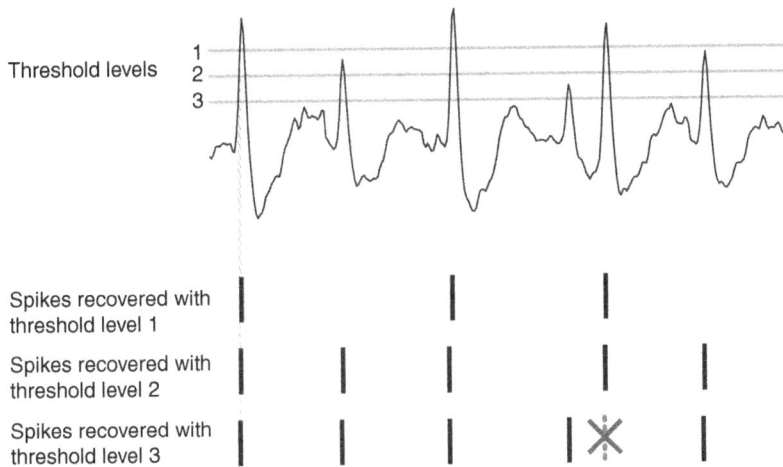

FIGURE 4.10 When selecting the threshold for spike detection, care must be taken to select a proper level. If the threshold level is too high, potentially isolatable neurons with smaller action potential waveforms will be missed. The negative effects of a threshold level that is too low are more subtle. The biggest problem comes from the fact that sampling a spike snippet blocks another spike from being detected for some amount of time. Thus, if the threshold level is too low, nonisolated neural activity can trigger acquisition that prevents immediately subsequent spikes that follow from being detected.

4. Start neural data acquisition software and ensure functionality.

5. Ensure there is enough disk space for neural and behavioral data.

6. Plug in animal.

7. Observe LFP and audio signal of reference electrode. Ensure no action potentials are apparent.

8. Observe and record signal quality on all channels. Especially watch for any day-to-day changes.

9. Set thresholds appropriately for each electrode.

10. Initiate saving and run experiment.

11. *(following experiment)* Close data files

12. Observe and record signal quality on all channels. Make small adjustments on electrodes—remember that slow drift will make the effect of these adjustments not apparent until 12–24 hours later.

13. Unplug animal and return to cage.

End-of-Experiment Procedures

Electrolytic Lesions

Following the conclusion of neural data acquisition, it is often important to assess the anatomical location of the recording electrodes. Preapplied dye can be used to deposit a record of the track of the electrode in the tissue (DiCarlo et al. 1996), but the best technique to discover the location of the electrical recording site is an electrolytic lesion.

These lesions are achieved by passing current through the recording electrodes, creating a small scar that can be seen later in histological sections.

To create an electrolytic lesion with tetrodes, an investigator needs a current generator (World Precision Instruments A365 or similar), a headstage connector breakout cable, and a timer. The following protocol can be followed:

1. Anesthetize the animal using an approved method.
2. Set the current level of the current generator to 30 µA. Set the current generator to "unipolar" current mode.
3. When areflexia is reached, connect the negative terminal of the current generator to the implant headstage connector position corresponding to the ground screw. Connect positive terminal to the first electrode for which a lesion is to be created. All channels of a tetrode can be connected or a single channel can be used. The experimenter should select channels that produced high signal quality, avoiding, for instance, broken or shorted electrode channels.
4. Activate the "On" setting of the current generator for 3–4 seconds. When current is flowing, the audio feedback tone will change pitch.
5. Move on to the next electrode.
6. Allow the animal to waken from anesthesia and return to cage. Animal can be euthanized immediately, but more robust lesions will be present if this is delayed for 24 hours.

Histology

Following electrolytic lesions, animals should be euthanized. For best results, follow euthanasia with transcardial perfusion of 4% paraformaldehyde. Electrodes may be retracted following perfusion, or the entire microdrive and brain can be extracted and postfixed for 24–48 hours. This later procedure often results in preservation of detectable electrode tracks, which can facilitate post hoc reconstruction of recording history in conjunction with electrolytic lesions. Following full fixation, the electrodes should be extracted and the brain prepared for sectioning. Typically 30 to 50 µm sections are used to locate electrolytic lesions in tissue stained for anatomical visualization either in bright field (e.g., cresyl violet) or in fluorescence (e.g., NeuroTrace, Life Technologies). An example section is shown in Figure 4.11.

Spike Sorting

Extracellular neural recordings nearly always record the activity from more than one neuron. Most of these neurons will have small indistinguishable action potential waveforms, but some will have unique signatures that can be consistently distinguished from those of the multi-unit mass and other single neurons. In the case of tetrode recordings, action potential waveforms need only be distinguishable in one of the four channels of the electrode. Consequently, in areas such as hippocampal area CA1 with densely packed neurons, >10 neurons can often be detected simultaneously with one tetrode.

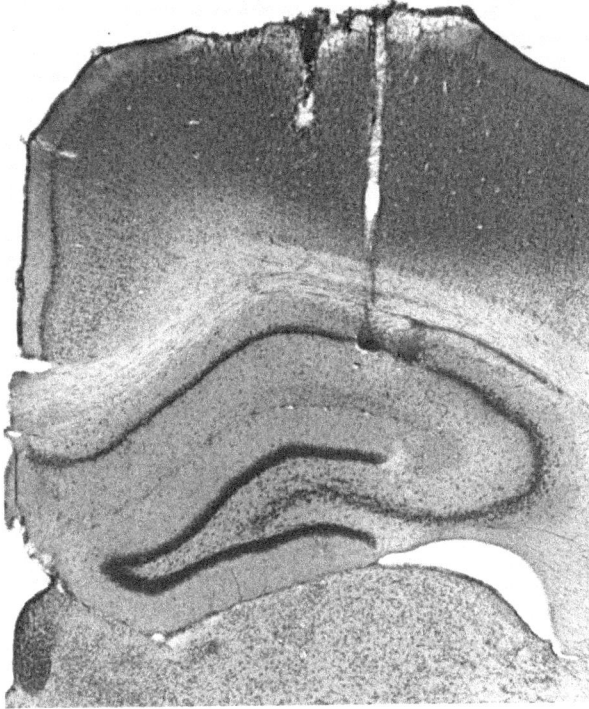

FIGURE 4.11 Sample histological section prepared using cresyl violet and imaged in brightfield illumination, showing electrode tracks and electrolytic lesions. (Photomicrograph by Shantanu Jadhav, used with permission.)

The process of disambiguating the spikes from multiple neurons is known as spike sorting. The problem of fully automated spike sorting has occupied engineers for some time (Carlson et al. 2014, Einevoll et al. 2012, Lewicki 1998, Wild et al. 2012). However, despite slow progress, most experimentalists will still use a manual or semiautomated approach for spike sorting. When spike snippets have been extracted during recording, spike sorting can begin immediately. Alternatively, following broadband recording, the acquired signal must first be filtered (typically 600 Hz to 6 kHz band-pass filter) and then snippets isolated. Some open-source packages such as KlustaSuite (http://klusta-team .github.io/) accomplish both the spike extraction and spike sorting tasks, whereas others such as MClust (http://redishlab.neuroscience.umn.edu/MClust/MClust.html) require spike snippets to be extracted first.

The waveforms of well-isolated single neurons will have features that are completely distinct from those of other neurons and stable across the time frame of a recording session. Some quantitative measures for the quality of isolation have been developed (Hill et al. 2011, Joshua et al. 2007, Schmitzer-Torbert et al. 2005). In cases where neurons are not well isolated, neural activity is often described as "multi-unit" and isolated clusters recorded on the same electrode are referred to as "units" rather than "neurons." In many cases, the shape of action potentials within a cluster of poorly isolated multi-unit activity may still provide information for further decoding or other analyses (Chestek et al. 2011, Ventura 2008, Wood & Black 2008, Zhiming et al. 2013).

Conclusion

Chronic recording from large ensembles of neurons has revolutionized systems neuroscience. Large ensemble recordings with excellent signal quality in rats and mice can be achieved using lab-constructed multi-tetrode microdrives. Many investigators find aspects of the process—construction of microdrives, elimination of noise in a recording environment, data processing, and so forth—initially quite challenging. With practice, however, trained experimentalists can initiate and conduct recording experiments and analyze the resulting data very rapidly. By carrying out these three aspects of an experiment in parallel in a staggered fashion, a complete experiment involving four to eight subjects can often be conducted in 9–12 months.

While we have described the steps in acquiring the signals of individual neurons in the context of chronic recordings in freely behaving animals, the systems and procedures have many similarities with those used in experiments involving head-fixed animals. These preparations often entail additional concerns about noise and, in the case of anesthetized subjects, maintenance of temperature and anesthetic plane. For acute experiments, where the electrodes are inserted during the recording session and subsequently removed, the microdrive can be replaced with larger, stereotactically positioned actuators. Furthermore, the inflammatory responses that can impair initial signal quality in chronic recording contexts typically do not present immediately, meaning that investigators can often record dozens of neurons daily using, for instance, multisite electrodes (Niell and Stryker 2010).

Careful design of behavioral paradigms coupled with high-quality recording using the techniques described above can yield experimental data that are so rich that subsequent analyses often produce new and important scientific understanding over many years (e.g., Mizuseki et al. 2014). Thus, while the experimental process can seem more tedious than other approaches in neuroscience that yield more immediate data, the profound understandings gained are more than worth the effort.

Protocol: 16-Tetrode Microdrive Construction

1. Overview

For chronic tetrode recording in behaving rats, an in-house–built microdrive array has a number of advantages over commercial ones—particularly its low cost (<$100) and customizability. The mechanism presented in this chapter was modified from designs used previously. Figure 4.12 shows a drawing of a "standard" 16-tetrode microdrive array as well as a version we built for the specialized context of simultaneous recording in multiple regions and simultaneous recording and stimulation.

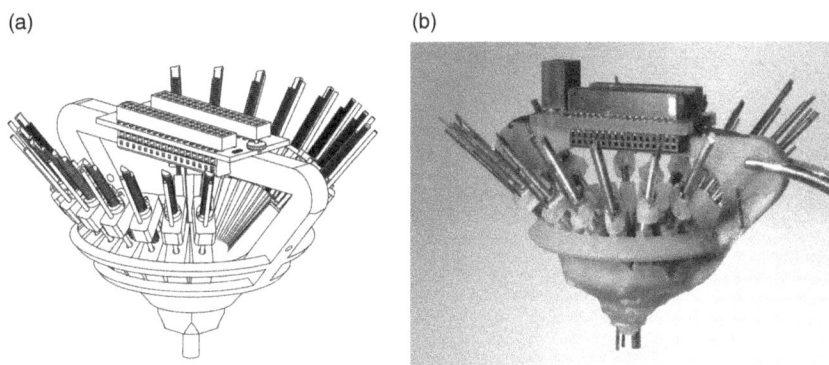

FIGURE 4.12 Drawing (a) and photograph (b) of a "standard" 16-tetrode microdrive array.

The design described below has evolved from one initially developed for a NASA mission (Knierim et al. 2000, Knierim and McNaughton 2001), with important contributions from Loren Frank and Mattias Karlsson (Frank et al. 2004, Karlsson et al. 2012, see also Lansink et al. 2007). Figure 4.13a shows the working design principle for the microdrive array. A portion of threaded rod is fixed in the 3D-printed (i.e., using a stereolithography process) body of the drive array. Each microdrive shuttle contains a circular nut that can turn freely but is captive (i.e., cannot translate within the shuttle). Using a toothed adjustment tool to turn the captive nut results in linear motion. With standard right-handed threaded rod, a right-handed rotation moves the shuttle downward. Using off-the-shelf 0–80 threaded rod, one turn corresponds to 318 μm of travel. A tetrode is glued to the shuttle within a polyimide sheath that itself is contained within a guide tube for mechanical stability. The polyimide sheath protects the tetrode at the entrance of the stainless-steel cannula that directs the tetrode to the bottom of the drive and from there to the brain. One or more assembling cannulae collect the tetrode cannulae into a bundle over the target regions of the brain and further align them vertically so that the tetrodes exit the device parallel to each other and perpendicular to the skull.

Preparing a microdrive array for implantation is a straightforward process. The main steps are making the shuttles, assembling and affixing the tetrode and gathering cannulae within the 3D-printed body, and making and loading tetrodes into the final assembly. Following an experiment, the tetrodes can be replaced, and the microdrive array can be sterilized (using ethylene oxide or other low-temperature means) and reused on a subsequent subject.

2. Needed supplies
   a. Stainless-steel tubing, 23 g, 30 g, and 14 g (Component Supply, Inc.), cut to length using a rotary tool. Small tubing should be cut using a diamond wheel.
   b. 5.5 or 6 mil wire (Component Supply, Inc.), cut to length with wire cutters or a rotary tool with a diamond wheel

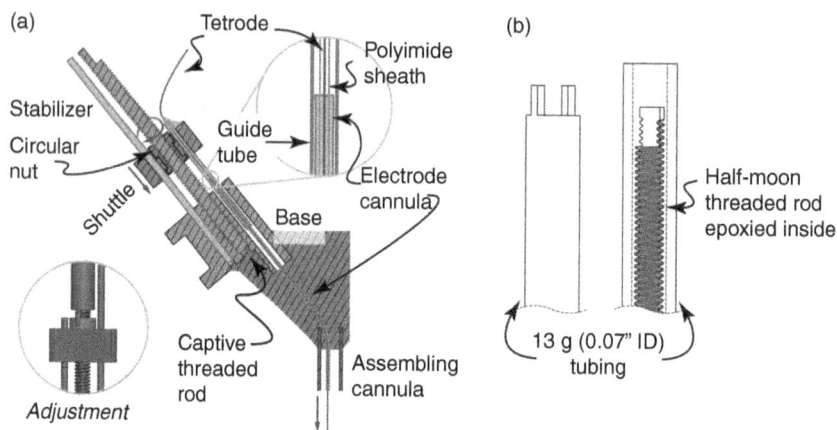

FIGURE 4.13 Schematic of a working design of a microdrive array.

    c. Threaded rod 0–80 (Component Supply, Inc.), cut to length and half-moon-carved using a rotary tool

    d. Polyimide tubing (Micro-Lumen, Inc.), cut to length with scissors.

    e. Methylmethacrylate powder and resin (Patterson Dental)

    f. Cyanoacrylate glue (normal or thick)

    g. Teflon lubricant

    h. Petroleum jelly mold release

    i. Custom-machined parts—slotted brass nuts, mold base; see designs online

    j. 3D-printed drive body, 3D-printed cannula mold (see designs online at http://github.com/kemerelab/MicrodriveDesigns)

    k. Electrode interface board (see designs online at http://github.com/kemerelab/MicrodriveDesigns), connectors

    l. Tetrode wire

    m. Tetrode heatshrink tubing, conductive paint

3. Needed tools

    a. Tetrode assembly station (Neuralynx, Inc.)

    b. NanoZ impedance measuring system (with adaptor) (Neuralynx, Inc.)

    c. 0–80 tap

    d. Cautery pen (Bovie)

    e. Acrylic mixing dish

    f. Belt sander

    g. Rotary tool with abrasive and diamond cutting wheel

    h. Slotted nut turn tool (make from 13 g tubing, see Fig. 4.13b)

    i. 0–80 turn tool (make from 13 g tubing and half-moon-cut threaded rod, see Fig. 4.13b)

4. Making the shuttles
   a. Begin by cutting needed tubing to length. This is a good time to additionally cut the 30 g tetrode cannulae and the assembling cannula that will be needed in future steps. The items needed for each of the 16 microdrives are shown in Figure 4.14. Note that microdrives with more complex targeting plans will require multiple assembling cannulas whose sizes are chosen appropriately.
   b. The shuttle mold is a square piece of brass (typically ~3 mm thick) with a centered 0–80 tapped hole flanked by holes drilled for 23-g tubing (0.75 mm separation).
   c. Install a pair of 23-g tubing, a 0–80 threaded rod, and a slotted nut onto the mold (Fig. 4.15a). This holds the slotted nut in the right orientation for acrylic to be added.
   d. Use a disposable bulb pipette to apply a drop of Teflon lubricant into the groove area of the slotted nut and onto the surface of the mold.
   e. Mix and apply methylmethacrylate cement. Apply a small quantity while cement is less viscous to ensure that no voids form between slotted nut and stabilizing rods (see Fig. 4.15b). As cement becomes more viscous, add more cement to form the appropriate ~3 mm height for the shuttle (see Fig. 4.15c).

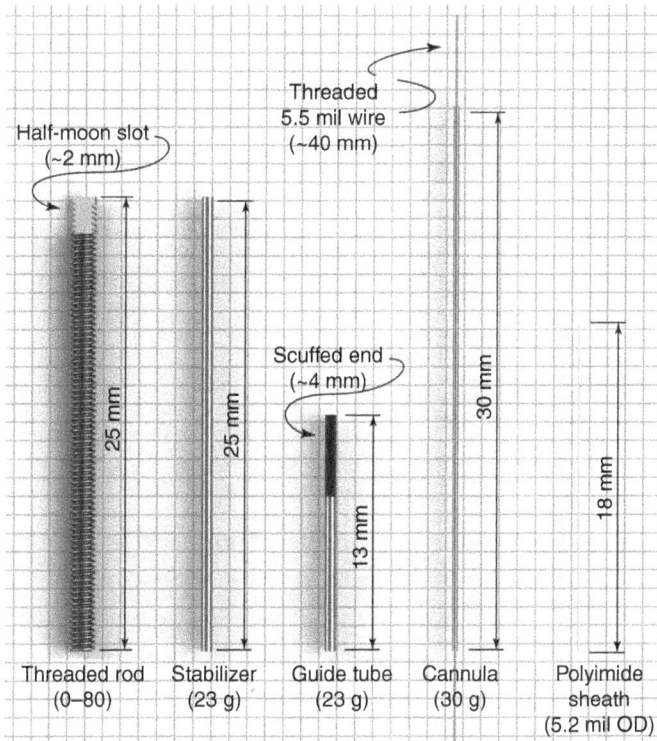

FIGURE 4.14 Items necessary to construct each microdrive.

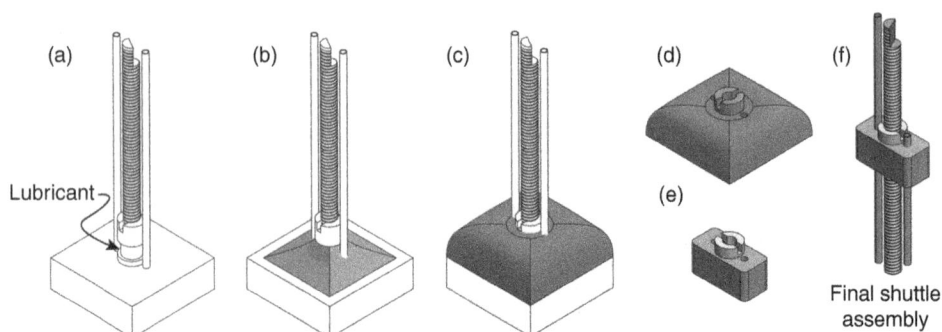

FIGURE 4.15 Construction of a microdrive shuttle. a: The shuttle mold with 23-g tubing, a 0–80 threaded rod, and a slotted nut. b, c: Application of methylmethacrylate cement to approximately 3 mm in height. d: Removal of unshaped shuttle from mold. e: Final shuttle after sanding. f: A completed shuttle.

    f. Remove tubing and threaded rod and use pliers to snap unshaped shuttle from mold (see Fig. 4.15d). Shape length and width of final shuttle on belt sander or rotary tool (see Fig. 4.15e). Test slotted nut for free rotation and discard if there is more than minimal friction or any wobble (friction or axial/lateral shifting will adversely affect final recording quality).

    g. Scuff the upper portion of 23 g guide tube using a file or rough sandpaper and glue into place in shuttles using cyanoacrylate (the glue will adhere more robustly to scuffed stainless steel). Take care not to let glue into interior of guide tube.

    h. Use a 23 g needle to ream the opposite hole in shuttle such that 23 g stabilizer translates freely. A completed shuttle assembly is shown in Figure 4.15f.

5. Assembling the drive body

    a. Use a 23 g needle to ream the channels for the 23 g guide tubes in the 3D-printed microdrive array base. Use a 0–80 tap to tap the holes for the threaded rod.

    b. Insert 5.5 mil wires into 30 g tetrode cannulae. This will prevent the cannulas from collapsing when being bent during assembly.

    c. Prepare the assembling cannula, scuffing the top ~2 mm of the outside.

    d. Insert all 16 stuffed 30 g cannulas into the assembling cannula. In the case of a more complex layout with multiple assembling cannulae, insert the proper number of 30 g cannulae into each.

    e. Mount this assembling cannula(e) into targeting block, and, as shown in Figure 4.16a use a clamp to support 3D-printed drive base above it.

    f. As shown in Figure 4.16b, carefully bend each 30 g cannula into an inner slot of the drive base and mount a shuttle over it (by inserting 23 g guide tube over the 30 g cannula, inserting 23 g stabilizer, and screwing threaded rod into microdrive array base). This will capture the 30 g cannula in place. Take care that the 30 g cannulas are not tangled as they leave the assembling cannula(e).

    g. Lower drive base to final height (see Fig. 4.16c). This will further bend the tubing, so be careful to avoid kinks.

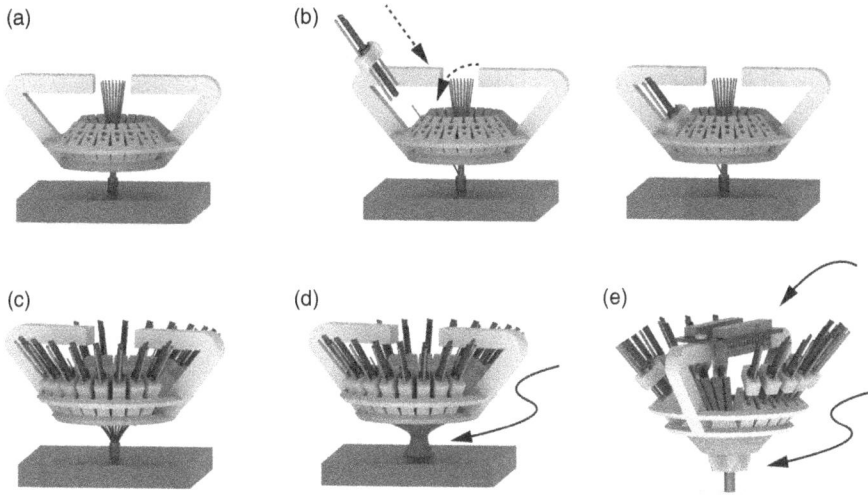

FIGURE 4.16 Assembling the drive body. a: The assembled cannulae is mounted into the targeting block. b: Each 30-g cannula is carefully bent into an inner slot of the drive base and a shuttle is mounted over it. c: The drive base is lowered to its final height. d: Methylmethacrylate is applied over the web of 30-g cannulae and interface. e: A rectangular bottom facilitates holding the drive during tetrode insertion.

h. Apply petroleum jelly to inner 23 g/30 g interface to prevent cement from entering and to targeting block surface to facilitate later removal.

i. Ensure that shuttles are lowered fully onto microdrive array base. As shown in Figure 4.16d, carefully apply methylmethacrylate cement over web of 30 g cannula and interface with assembling cannula and microdrive. Note that the (lowered) placement of each guide tube will now constrain the maximum depth to which each shuttle can be lowered.

j. After cement has hardened (typically ~20 minutes, although full strength is not reached for up to 24 hours), remove assembly from targeting block. Shape applied cement with a belt sander or rotary tool to minimize weight. A rectangular bottom facilitates holding the drive during tetrode insertion (see Fig. 4.16e).

k. Fill in interstitial spaces in bottom of assembly cannula by deeply inserting stainless-steel wire with pliers. This ensures that fluid cannot enter and remain after implants.

l. Carefully cut the assembly cannula from the bottom, together with the 30 g cannulas and any added wire fill, using a rotary tool with a diamond cutting wheel. It is critical that the 30 g cannulas still be stuffed with 5.5 mil wire during cutting to prevent permanent stoppages.

m. Cut the top of the 30 g tubing to final length and clean the inner portion of the opening using the tip of a needle. Ensure that 5.5 mil wire still moves smoothly through each 30 g cannula.

n. Record the pattern of cannulas at the bottom of the assembling cannula. Note which microdrive corresponds to which bottom position. In a tightly packed assembling cannula, adjacent tetrodes will protrude with about 125 μm of separation. Having a record of the pattern will facilitate understanding where each tetrode is within the brain.

o. Mount an electrode interface board (EIB) to the top of the microdrive array using small screws and/or epoxy. In the latter case, be careful not to allow contamination of the electrical contacts.

6. Loading tetrodes

Below is a summary of one process for loading tetrodes, which involves attaching each channel of the tetrode to the male pin of a connector. Other approaches are also popular, such as the use of a small gold pin or solder to affix each wire into a hole in the EIB. Most of the process can be performed using magnifying goggles, but the use of a stereomicroscope is strongly recommended.

a. Make tetrodes. See Nguyen et al. (2009) for details.

b. Cut the four wires on top of the tetrode to length (2–6 mm).

c. Gently rub the hot tip of a cautery tool along the final ~2 mm of each wire. This will strip the insulation.

d. Cut small pieces of 50-μm-diameter heatshrink tubing to ~2 mm in length. Mount a four-pin connector in a clamp in the focal zone of the microscope. Put a drop of conductive silver paint (Silver Print II, GC Electronics) onto a nonporous surface. Using tweezers, carefully dip the end of a tubing segment in the silver paint so that capillary action causes it to half-fill.

e. Grasping the tetrode in your fingers, use tweezers to place a dipped tubing segment on each of the four wires at the end of the tetrode.

f. Carefully use tweezers to place each tubing segment onto one of the pins of the connector. As shown in Figure 4.17a, the tubing segments should be oriented so that the junction of four tetrode wires is adjacent to the body of the connector. Ensure that the portion of the tetrode's wire with insulation removed is in contact with the connector pin inside each tubing segment. Using a heat gun, shrink the tubing segments, attaching the tetrodes to the four-pin connector.

g. Load a polyimide sheath into a microdrive shuttle such that it protrudes ~1 mm (see Fig. 4.17b). This will require that the polyimide sheath also is contained within the 30 g cannula, as shown in Figure 4.13.

h. Plug four-pin connector into the EIB, and use carbon-tipped forceps to carefully insert tetrode into the polyimide sheath/shuttle/drive (see Fig. 4.17c).

i. Using the electroplating setup, test the tetrode for shorts and continuity. Remove and replace if any tetrode channels are shorted together or not connected. As this requires plugging into the EIB, it is sometimes easiest to test a group of tetrodes all at once.

j. Using a small piece of scrap wire or other tool, apply cyanoacrylate glue to attach the tetrode, the polyimide sheath, the 23-g guide tube, and the shuttle. Be careful to avoid adhering the grooved nut to the body of the shuttle.

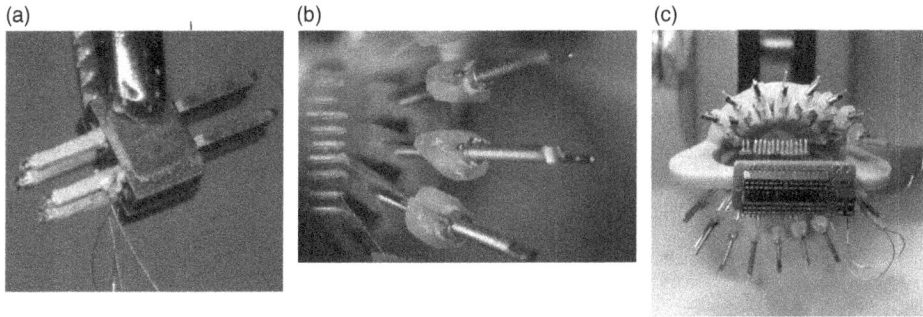

FIGURE 4.17 Loading tetrodes. a: The tubing segments should be oriented so that the junction of four tetrode wires is adjacent to the body of the connector. b: A polyimide sheath is loaded into a microdrive shuttle such that it protrudes ~1 mm. c: The tetrode is carefully inserted into the polyimide sheath/shuttle/drive.

7. Plating
   a. Carefully cut the protruding end of each tetrode to appropriate length. Typically, this corresponds to the full travel distance of the corresponding microdrive such that the tip of the tetrode retracts just within its cannula when it is fully retracted.
   b. Measure preplating impedance in saline. Check again for short-circuited connections and nonconnected channels.
   c. Electroplate tetrodes. Small plating currents (0.2–0.5 μA flowing from plating solution to tetrode for ~0.5 s) will result in a slower but well-controlled process.
   d. Rinse in distilled water and measure impedance in saline. Repeat plating, rinsing, and measuring until desired final impedance (typically 175–250 kOhms at 1 kHz) is reached and is stable after 1 hour.
   e. Expect two to three rounds of plating. Higher currents and very low impedances will often correspond with short-circuited connections between the channels of the tetrode (although modifications of the plating solution can achieve lower impedances; Ferguson et al. 2009).
   f. Retract tetrodes into drive for presurgical sterilization.
8. After the experiment
   a. Remove old tetrodes and attached shuttles by unplugging from EIB. Using needle-nose pliers, grasp and twist each 23-g guide tube to remove from shuttles. Guide tubes can be soaked in acetone to dissolve cyanoacrylate glue and remove polyimide tubing, or discarded.
   b. Carefully remove skull fragments and methylmethacrylate cement added to microdrive array during implantation to reveal bottom 3–4 mm of assembling cannula. Most of the cement should be removed using a rotary tool. Use flush cutter shears to carefully fracture and remove the remaining cement to reveal the undamaged assembling cannula. With care, the microdrive array can be returned to a preimplantation state and used many times.

Frequently Asked Questions

Q: How can I manually measure the impedance of tetrodes? How can I manually electroplate tetrodes?

A: Rather than using a multiplexing impedance tester/electroplater, some investigators choose to measure the impedance of each channel of each electrode independently. Devices made for this purpose (e.g., IMP-2, Bak Electronics, Inc., Sanford, FL) emit a very-low-current sine wave and measure the resulting voltage, allowing for impedance measurement up to several megaohms. These devices can be used in conjunction with a current generator (e.g., A365, World Precision Instruments, Inc., Sarasota, FL) to electroplate a single electrode to a desired level of impedance.

Q: When starting data acquisition, what parameters should I use for the filter used for detecting spikes? What is a good beginning value for the spike detection threshold?

A: Most data acquisition systems use a passband filter on incoming neural activity to remove low-frequency LFP signals and high-frequency noise before performing spike detection. A good initial setting for this passband is 300–6,000 Hz. If there are many neurons with large action potentials, using a slightly narrower band (e.g., 600–6,000 Hz) may improve the separability. For electrodes with ~200-kΩ impedance, a good initial value for the spike threshold is 40 μV. As described in the text, in the presence of large action potentials, it is often advisable to increase to 60 μV or more.

Q: How do I pick a reference electrode? When should I move my reference electrode?

A: Investigators employ two common strategies for choosing a local reference electrode for postsampling noise reduction (see text). One strategy is to pick an electrode postimplantation that has poor signal characteristics (e.g., a bad channel). The advantage of this approach is the fact that maximal use is made of the microdrives in an array. Disadvantages include the fact that the chosen reference electrode will typically have impedance characteristics that allow it to sense action potentials, the horizontal targeting may not be ideally located for a target reference area, and it can sometimes be challenging to know which tetrode will be "bad" on the first few days of recording. One strategy is to implant a dedicated electrode with lower impedance than typically used for neural recording. The advantage of this strategy is that the signal properties are optimized for referencing. The disadvantage is that one or more of the microdrives in a microdrive array must be dedicated for the purposes of referencing. Thus, a dedicated reference is appropriate when a region with little neural activity (e.g., a white matter tract) is not present in the vicinity of the recording electrodes.

In general, the reference electrode should be placed somewhere with "noise" characteristics (e.g., artifacts due to jaw muscle movements) in common with the data electrodes and no "signal" (i.e., neural activity). The second criterion can be partially achieved by locating the reference electrode in a white matter tract such as the commissural fibers below cortex, as action potentials are rarely detected near axons. Note that lower-frequency LFP signal is still present in these regions, and thus any analysis must either be unreferenced or otherwise account for the location of the reference electrode. In general, the reference electrode should be moved so as to keep the reference signal free of action potentials. This sometimes requires multiple adjustments in the first few weeks of a recording experiment.

Q: What amplitudes should I expect of well-isolated action potentials? How can I tell if I'm recording from one or multiple neurons?

A: A well-isolated neuron typically displays action potentials in the range of several hundred μV. Signals >1 mV are observed on occasion. In addition to amplitude, a well-isolated neuron will also display stability in action potential waveform shape, with variability comparable to the level of background noise. Careful electrode adjustment should result in neurons whose action potentials are stable over many hours or even multiple days. Any analysis should be predicated on the shape of the action potential waveform having been stable over the period analyzed. Instability can often be seen simply by plotting the peak amplitude of the action potential waveform as a function of time. Note that in bursting neurons, the situation is more complex, as the shape of action potentials in bursts *does* decrease in amplitude. In most scenarios with action potentials from multiple neurons detectable on the same electrode(s), the action potential shape or amplitude will be clearly different. A key test, however, is to evaluate whether the interspike interval of a putatively isolated unit shows violations of a biophysical refractory period (1–2 ms). Any violations strongly suggest the presence of multiple neurons in the sample.

Q: In what increment should I move the electrode to find better activity? How do I adjust microdrives in an awake animal?

A: The adjustment process described in the text addresses reaching the target region, using electrophysiological signs and the recorded depth of the electrode relative to an atlas or other experimentally derived depth map. Once the electrode is in the vicinity of the desired neurons, however, adjustment should proceed with minimal electrode steps. A minimal step will result in a small but noticeable change in the audio monitor signal, and sometimes even these minimal steps should be counterbalanced by retractions (i.e., 40–60 μm forward followed by 20–30 μm backward). High-quality isolations cannot be assessed until the following recording session as a result of slow relaxations of the brain tissue. Adjusting a microdrive in a moving animal is a challenging process. Addressing the needs of the animal—food, water, or safety in the form of a dark corner—may facilitate adjustment. Carrying out the process during the rest phase of the animal's diurnal cycle may promote sleep that can aid the process. In general, minimal force should be applied to the implant to prevent loosening of the attaching screws and adhesive. Thus, one strategy is to grasp the base of the implant with the nondominant hand and move along with the animal in the adjustment environment. The adjustment tool in the dominant hand also moves laterally but only rotates when the animal's movement pauses.

Q: How do I know where I am in the brain? How can I tell different types of neurons apart?

A: One of the biggest challenges for a new investigator is to develop an intuition for the different characteristics of relevant brain regions (i.e., the target and the regions above, below, and in close proximity). Some information may be gleaned from atlases (e.g., the presence of a white matter tract) or from the literature (e.g., stereotypical patterns of activity or relevant receptive fields). The presence of characteristic action potentials from pyramidal cells (broad, asymmetric waveforms [Sirota et al. 2008]), fast spiking interneurons (narrow waveforms), or axons (very symmetric or triphasic waveforms) may often be enlightening. However, careful note taking *during the adjustment process* in the first few subjects and postexperiment lesions and histological verification are critical.

Q: What might cause all of my signal to suddenly degrade into noise?

A: Don't panic! Whenever one sees unusually large amounts of noise, it's always best to check the physical connections in the system. Are headstages plugged in securely with the proper orientation? Does signal from an artificial signal generator work? Sometimes static buildup in the recording environment can lead to bad noise characteristics, which can be alleviated with an antistatic spray. In the worst case scenario, if the data acquisition signal pathway is intact, a sudden change in recording quality can sometimes indicate that the microdrive array may be separating from the skull, so extra care should be taken.

References

Averbeck, B. B., Latham, P. E., & Pouget, A. (2006). Neural correlations, population coding and computation. *Nature Rev Neurosci* 7(5): 358–366.

Averbeck, B. B., & Lee, D. (2004). Coding and transmission of information by neural ensembles. *Trends Neurosci* 27(4): 225–230.

Bragin, A., Jandó, G., Nádasdy, Z., et al. (1995). Gamma (40–100 Hz) oscillation in the hippocampus of the behaving rat. *J Neurosci* 15(1 Pt 1): 47–60.

Briggman, K. L., Abarbanel, H. D. I., & Kristan, W. B. (2006). From crawling to cognition: analyzing the dynamical interactions among populations of neurons. *Curr Opin Neurobiol* 16(2): 135–144.

Buzsáki, G. (2004). Large-scale recording of neuronal ensembles. *Nature Neurosci* 7(5): 446–451.

Buzsáki, G., Horváth, Z., Urioste, R., et al. (1992). High-frequency network oscillation in the hippocampus. *Science* 256(5059): 1025–1027.

Carlson, D. E., Vogelstein, J. T., Wu, Q., et al. (2014). Multichannel electrophysiological spike sorting via joint dictionary learning and mixture modeling. *IEEE Trans Biomed Engin* 61(1): 41–54.

Chalk, M., Herrero, J. L., Gieselmann, M., et al. (2010). Attention reduces stimulus-driven gamma frequency oscillations and spike field coherence in V1. *Neuron* 66(1): 114–125.

Chen, B., Carlson, D. E., Carin, L., et al. (2011). On the analysis of multi-channel neural spike data. In J. Shawe-Taylor, R. S. Zemel, P. Bartlett, et al. (Eds.), *Advances in neural information processing systems 24* (pp. 936–944).

Chestek, C., Gilja, V., Nuyujukian, P., et al. (2011). Long-term stability of neural prosthetic control signals from silicon cortical arrays in rhesus macaque motor cortex. *J Neural Engin* 8(4): 045005.

Churchland, M. M., Yu, B. M., Sahani, M., & Shenoy, K. V. (2007). Techniques for extracting single-trial activity patterns from large-scale neural recordings. *Curr Opin Neurobiol* 17(5): 609–618.

Cogan, S. F. (2008). Neural stimulation and recording electrodes. *Ann Rev Biomed Engin* 10: 275–309.

DiCarlo, J. J., Lane, J. W., Hsiao, S. S., & Johnson, K. O. (1996). Marking microelectrode penetrations with fluorescent dyes. *J Neurosci Meth* 64(1): 75–81.

Drake, K. L., Wise, K. D., Farraye, J., et al. (1988). Performance of planar multisite microprobes in recording extracellular single-unit intracortical activity. *IEEE Trans Biomed Engin* 35(9): 719–732.

Ecker, A. S., Berens, P., Keliris, G., et al. (2010). Decorrelated neuronal firing in cortical microcircuits. *Science* 327(5965): 584–587.

Einevoll, G. T., Franke, F., Hagen, E., et al. (2012). Towards reliable spike-train recordings from thousands of neurons with multielectrodes. *Curr Opin Neurobiol* 22(1): 11–17.

Ferguson, J. E., Boldt, C., & Redish, A. D. (2009). Creating low-impedance tetrodes by electroplating with additives. *Sensors and Actuators. A, Physical* 156(2): 388–393.

Frank, L. M., Stanley, G. B., & Brown, E. N. (2004). Hippocampal plasticity across multiple days of exposure to novel environments. *J Neurosci* 24(35): 7681–7689.

Fraser, G. W., & Schwartz, A. B. (2012). Recording from the same neurons chronically in motor cortex. *J Neurophysiol* 107(7): 1970–1978.

Goard, M., & Dan, Y. (2009). Basal forebrain activation enhances cortical coding of natural scenes. *Nature Neurosci* 12(11): 1444–1449.

Gray, C. M., Maldonado, P. E., Wilson, M., & McNaughton, B. (1995). Tetrodes markedly improve the reliability and yield of multiple single-unit isolation from multi-unit recordings in cat striate cortex. *J Neurosci Meth* 63(1-2): 43–54.

Harris, K. D. (2005). Neural signatures of cell assembly organization. *Nature Rev Neurosci* 6(5): 399–407.

Harris, K. D., Henze, D. A., Csicsvari, J., et al. (2000). Accuracy of tetrode spike separation as determined by simultaneous intracellular and extracellular measurements. *J Neurophysiol* 84(1): 401–414.

Harris, K. D., & Mrsic-Flogel, T. D. (2013). Cortical connectivity and sensory coding. *Nature* 503(7474): 51–58.

Harris, K. D., & Thiele, A. (2011). Cortical state and attention. *Nature Rev Neurosci* 12(9): 509–523.

Harrison, R. R. (2008). The design of integrated circuits to observe brain activity. *Proc IEEE* 96(7): 1203–1216.

Henze, D., Borhegyi, Z., Csicsvari, J., et al. (2000). Intracellular features predicted by extracellular recordings in the hippocampus in vivo. *J Neurophysiol* 84(1): 390–400.

Hill, D. N., Mehta, S. B., & Kleinfeld, D. (2011). Quality metrics to accompany spike sorting of extracellular signals. *J Neurosci* 31(24): 8699–8705.

Jog, M. S., Connolly, C. I., Kubota, Y., et al. (2002). Tetrode technology: advances in implantable hardware, neuroimaging, and data analysis techniques. *J Neurosci Meth* 117(2): 141–152.

Joshua, M., Elias, S., Levine, O., & Bergman, H. (2007). Quantifying the isolation quality of extracellularly recorded action potentials. *J Neurosci Meth* 163(2): 267–282.

Karlsson, M. P., Tervo, D. G. R., & Karpova, A. Y. (2012). Network resets in medial prefrontal cortex mark the onset of behavioral uncertainty. *Science* 338(6103): 135–139.

Kass, R. E., Ventura, V., & Brown, E. N. (2005). Statistical issues in the analysis of neuronal data. *J Neurophysiol* 94(1): 8–25.

Kentros, C., Hargreaves, E., Hawkins, R. D., et al. (1998). Abolition of long-term stability of new hippocampal place cell maps by NMDA receptor blockade. *Science* 280(5372): 2121–2126.

Kloosterman, F., Davidson, T. J., Gomperts, S. N., et al. (2009). Micro-drive array for chronic in vivo recording: drive fabrication. *J Vis Exp* 26: 2–5.

Knierim, J. J., & McNaughton, B. L. (2001). Hippocampal place-cell firing during movement in three-dimensional space. *J Neurophysiol* 85(1): 105–116.

Knierim, J. J., McNaughton, B. L., & Poe, G. R. (2000). Three-dimensional spatial selectivity of hippocampal neurons during space flight. *Nature Neurosci* 3(3): 209–210.

Lansink, C. S., Bakker, M., Buster, W., et al. (2007). A split microdrive for simultaneous multi-electrode recordings from two brain areas in awake small animals. *J Neurosci Methods* 162, 129–138.

Lee, S.-H., & Dan, Y. (2012). Neuromodulation of brain states. *Neuron* 76(1): 209–222.

Lewicki, M. S. (1998). A review of methods for spike sorting: the detection and classification of neural action potentials. *Network (Bristol, England)* 9(4): R53–78.

McNaughton, B. L., O'Keefe, J., & Barnes, C. A. (1983). The stereotrode: a new technique for simultaneous isolation of several single units in the central nervous system from multiple unit records. *J Neurosci Meth* 8(4): 391–397.

Mizuseki, K., Diba, K., Pastalkova, E., et al. (2014, May). Neurosharing: large-scale data sets (spike, LFP) recorded from the hippocampal-entorhinal system in behaving rats. *F1000 Research* 1–14.

Mizuseki, K., Royer, S., Diba, K., & Buzsáki, G. (2012). Activity dynamics and behavioral correlates of CA3 and CA1 hippocampal pyramidal neurons. *Hippocampus* 22(8): 1659–1680.

Mizuseki, K., Sirota, A., Pastalkova, E., & Buzsáki G. (2009). Theta oscillations provide temporal windows for local circuit computation in the entorhinal-hippocampal loop. *Neuron* 64(2): 267–380.

Nguyen, D. P., Layton, S. P., Hale, G., et al. (2009). Micro-drive array for chronic in vivo recording: tetrode assembly. *J Vis Exp* 26: pii: 1098. doi: 10.3791/1098.

Nicolelis, M. A. L., Dimitrov, D., Carmena, J. M., et al. (2003). Chronic, multisite, multielectrode recordings in macaque monkeys. *Proc Nat Acad Sci USA* 100(19): 11041–11046.

Niell, C. M., & Stryker, M. P. (2010). Modulation of visual responses by behavioral state in mouse visual cortex. *Neuron* 65(4): 472–479.

Normann, R. A., Campbell, P. K., & Jones, K. E. (1991). *Micromachined, silicon-based electrode arrays for electrical stimulation of or recording from cerebral cortex*. Presented at Conference: Micro Electro Mechanical Systems, 1991, MEMS '91, Proceedings. An Investigation of Micro Structures, Sensors, Actuators, Machines and Robots, pp. 247–252.

Polikov, V. S., Tresco, P. A., & Reichert, W. M. (2005). Response of brain tissue to chronically implanted neural electrodes. *J Neurosci Meth* 148(1): 1–18.

Ponce-Alvarez, A., Nacher, V., Luna, R., et al. (2012). Dynamics of cortical neuronal ensembles transit from decision making to storage for later report. *J Neurosci* 32(35): 11956–11969.

Schmitzer-Torbert, N., Jackson, J., Henze, D., Harris, K., & Redish, A. D. (2005). Quantitative measures of cluster quality for use in extracellular recordings. *Neuroscience* 131(1): 1–11.

Sirota, A., Montgomery, S., Fujisawa, S., et al. (2008). Entrainment of neocortical neurons and gamma oscillations by the hippocampal theta rhythm. *Neuron* 60(4): 683–697.

Stevenson, I. H., & Kording, K. P. (2011). How advances in neural recording affect data analysis. *Nature Neurosci* 14(2): 139–142.

Tolias, A. S., Ecker, A. S., Siapas, A. G., et al. (2007). Recording chronically from the same neurons in awake, behaving primates. *J Neurophysiol* 98(6): 3780–3790.

Vandecasteele, M. S., Royer, S., Belluscio, M., et al. (2012). Large-scale recording of neurons by movable silicon probes in behaving rodents. *J Vis Exp* 61: e3568.

Ventura, V. (2008). Spike train decoding without spike sorting. *Neural Computation* 20(4): 923–963.

Vetter, R. J., Williams, J. C., Hetke, J. F., et al. (2004). Chronic neural recording using silicon-substrate microelectrode arrays implanted in cerebral cortex. *IEEE Trans Biomed Engin* 51(6): 896–904.

Voigts, J., Siegle, J. H., Pritchett, D. L., & Moore, C. I. (2013). The flexDrive: an ultra-light implant for optical control and highly parallel chronic recording of neuronal ensembles in freely moving mice. *Frontiers in Systems Neuroscience* 7(May): 8.

Wild, J., Prekopcsak, Z., Sieger, T., et al. (2012). Performance comparison of extracellular spike sorting algorithms for single-channel recordings. *J Neurosci Meth* 203(2): 369–376.

Wilson, M. A., & McNaughton, B. L. (1993). Dynamics of the hippocampal ensemble code for space. *Science* 261(5124): 1055–1058.

Wood, F., & Black, M. J. (2008). A nonparametric Bayesian alternative to spike sorting. *J Neurosci Meth* 173(1): 1–12.

Ylinen, A., Bragin, A., Nádasdy, Z., et al. (1995). Sharp wave-associated high-frequency oscillation (200 Hz) in the intact hippocampus: network and intracellular mechanisms. *J Neurosci* 15(1 Pt 1): 30–46.

Zhiming, X., Keng, A. K., & Cuntai, G. (2013). *Neural decoding of movement targets by unsorted spike trains*. Presented at 2013 IEEE International Conference on Acoustics, Speech and Signal Processing (ICASSP), pp. 954–958.

Fast-Scan Cyclic Voltammetry in Behaving Animals

Monica M. Arnold, Lauren M. Burgeno, and
Paul E. M. Phillips

Introduction

Chemical neurotransmission is a process by which neurons and neural circuits convey information within the nervous system via release of neurotransmitters from presynaptic neurons onto receptors of postsynaptic neurons.

Patterns of synaptic transmission include continuous release of neurotransmitters that provide tone at receptors, driven by low-frequency neural activity, interspersed with transient increases in the extracellular neurotransmitter concentration, driven by short bursts of activity. These patterns of transmission are often termed tonic or phasic neurotransmission, respectively. Each of these signaling modalities is hypothesized to mediate discrete types of information transfer, and thus they each may serve different computational roles. Central monoamine systems, including circuits containing dopamine-, norepinephrine- and serotonin-releasing neurons, all exhibit both tonic and phasic signaling. For example, hypotheses on the role of tonic and phasic dopamine neurotransmission from the substantia nigra/ventral tegmental area include regulation of response vigor, and updating values assigned to reward-related stimuli and actions, respectively (Montague et al. 1996, Niv 2007). Norepinephrine neurons in the locus coeruleus exhibit phasic and tonic signaling that has been associated with focused attention and disengagement or distractibility, respectively (Aston-Jones and Cohen 2005). The significance of dorsal raphe serotonergic-tonic and phasic firing remains to be fully elucidated (Jennings 2013), although there is evidence that serotonergic dorsal raphe neurons display firing patterns that synchronize with the organism's sleep–wake cycle and environmental stimuli (Jacobs and Fornal 1991). Given the diversity of behavioral function attributed to these different patterns of transmission, it is important to use detection methods that can parse tonic and phasic signaling modalities in awake, freely moving animals throughout the course of behavioral tasks.

While traditional methods for assessing electrical neurotransmission (i.e., electrophysiology) can detect both phasic and tonic activity, traditional methods for assessing chemical transmission (e.g., microdialysis) do not afford this benefit due to the low temporal resolution of sampling that involves harvesting material from the brain for offline analysis. *In vivo* electrochemical approaches (e.g., constant-potential amperometry, chronoamperometry, and fast-scan cyclic voltammetry) circumvent this problem by making in situ measurements directly in the brain. These methodologies detect the concentration of molecules of interest by measuring the electrical currents generated by their electrolysis following application of an electrical potential. Therefore, their use is restricted to the detection of neurotransmitters that readily undergo oxidation–reduction (redox) reactions, such as monoamines. Nonetheless, these techniques offer the promise of resolving time-dependent changes in dopamine, norepinephrine, serotonin, histamine, and adenosine during behavioral tasks. Thus far, all of these electroactive neurotransmitters have been successfully measured using electrochemical approaches in simplified biological preparations (Brazell et al. 1987, Ewing et al. 1982, Palij and Stamford 1992, Pihel et al. 1996, Swamy and Venton 2007). However, to date, almost all of the studies characterizing behaviorally evoked changes in neurotransmission using *in vivo* electrochemistry have been limited to dopamine. Those studied have typically utilized fast-scan cyclic voltammetry (FSCV), which has the highest chemical resolution of the *in vivo* electrochemical methods commonly employed in neuroscience (Phillips and Wightman 2003) and can detect subsecond changes in extracellular dopamine time-locked to specific behavioral events (Phillips et al. 2003a). Therefore, in this chapter, we focus on the methodological practicalities that have enabled FSCV as a tool for studying the critical nuances of dopamine neurotransmission that underlie cognition and behavior.

Fundamentals of Fast-Scan Cyclic Voltammetry

Several biologically important molecules are electrochemically active, including the monoamine neurotransmitters norepinephrine, serotonin and dopamine, which all can undergo oxidation–reduction or redox reactions (i.e., they undergo a loss and gain of electrons that depends on the surrounding voltage). The range of electrical potentials at which a molecule will undergo a redox reaction depends on its molecular structure. In the case of dopamine, when it becomes oxidized, it loses two electrons to become dopamine-o-quinone (Fig. 5.1). Under the appropriate conditions, dopamine-o-quinone readily converts back to dopamine by gaining two electrons via a reduction reaction. This chemical change can be induced in molecules on the surface of an electrode by applying an electrical potential to the electrode. The flow of electrons between the dopamine molecule and the electrode during this reaction can be measured as faradaic current. This current is directly proportional to the concentration of dopamine molecules that were oxidized, yielding a method for qualifying the dopamine concentration at the surface of the electrode.

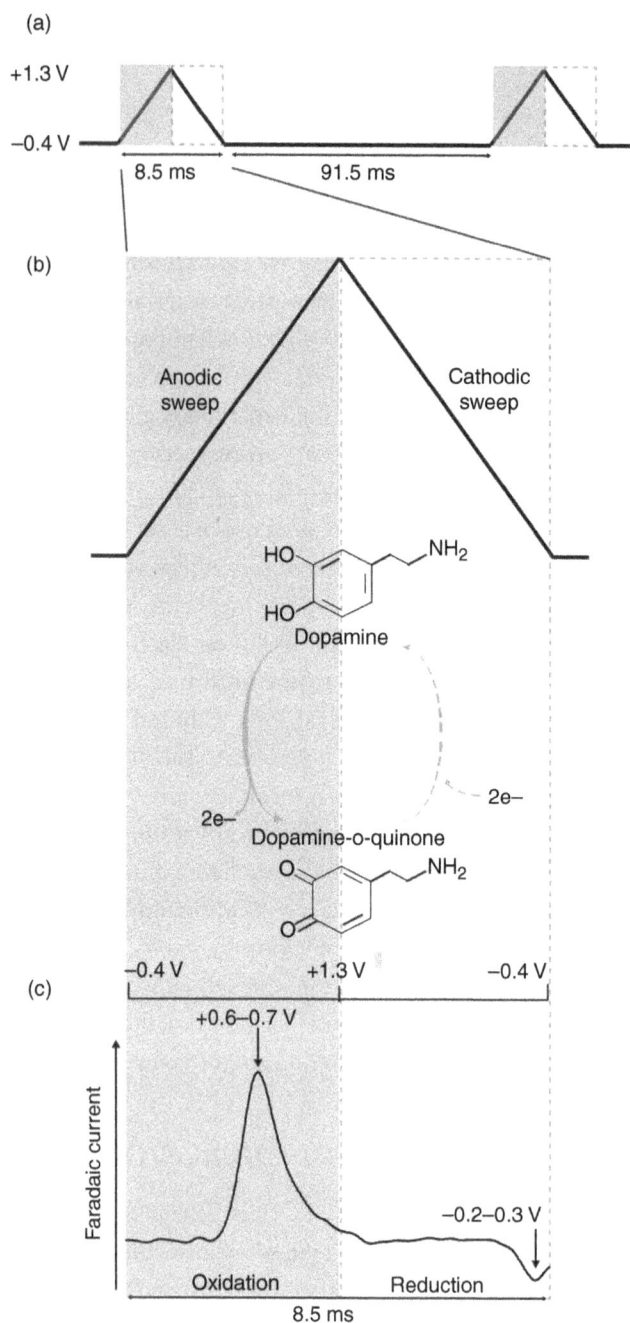

(a)

+1.3 V

−0.4 V

8.5 ms   91.5 ms

(b)

Anodic sweep

Cathodic sweep

HO
HO
Dopamine
NH$_2$

2e−   2e−

Dopamine-o-quinone

O
O
NH$_2$

−0.4 V   +1.3 V   −0.4 V

(c)

+0.6–0.7 V

Faradaic current

−0.2–0.3 V

Oxidation   Reduction

8.5 ms

FIGURE 5.1 FSCV waveform and the electrochemical signature of dopamine. A: The triangular FSCV waveform consists of a linear upward voltage sweep (anodic sweep) from −0.4 V to +1.3 V and then subsequent linear downward voltage sweep (cathodic sweep) from +1.3 V to −0.4 V, all over the course of 8.5 ms. This waveform is applied once every 100 ms and between applications the voltage is held at −0.4 V. B: The application of this triangular waveform induces the oxidation of dopamine to dopamine-o-quinone during the anodic sweep via liberation of two electrons, and then subsequent reduction of any dopamine-o-quinone remaining at the electrode surface back to dopamine via gain of two electrons during the cathodic sweep. C: The flow of electrons generated during these reactions is measured as faradaic current at the carbon fiber microelectrode. The peak dopamine oxidation current is typically observed at +0.6 to +0.7 V and peak reduction current at −0.2 to −0.3 V. This can be presented as a time course (i.e., current vs. time plot) or as a CV (current vs. voltage plot, C inset). In the inset, the oxidation current measured during the anodic sweep is drawn as a solid gray line and the reduction current measured during the cathodic sweep as a black dashed line.

For electrochemical detection of dopamine, a voltage is applied between a working and a reference electrode. The working electrode is made with an electrically conductive material that can readily detect the transfer of electrons to and from dopamine on the surface of the electrode but does not undergo redox reactions itself. Bare carbon-fiber working electrodes are the conventional choice for biological applications; however, Nafion-coated electrodes are sometimes used to improve analyte selectivity in the presence of interfering anions (Gerhardt et al. 1984). A standard silver/silver chloride (Ag/AgCl) reference electrode is used to maintain the absolute potential at the working electrode. The potential is applied to the electrode as a triangular waveform that is linearly ramped from the initial holding potential (typically –0.4 V vs. Ag/AgCl) to a maximum voltage (typically +1.3 V vs. Ag/AgCl) and then returned to the holding potential (see Fig. 5.1A). For the applications we will describe, this voltage scan is completed in 8.5 ms, yielding a scan rate of 400 V/s (3.4 V [total voltage sweep]/0.0085 s = 400 V/s). The potential is then maintained at the holding potential until the next voltage scan. Scans are typically applied every 100 ms (10-Hz sampling; see Fig. 5.1A). The faradaic current (electron loss and gain) generated from oxidation and reduction of dopamine molecules at the electrode surface during the triangular waveform application (see Fig. 5.1B) can be depicted as a cyclic voltammogram (CV), or a current-voltage (IV) curve, which serves as an electrochemical signature characteristic of the analyte of interest (see Fig. 5.1C). Since FSCV allows for the collection of ten voltammetric scans per second, color plots are often used to depict the changes in current generated at each point in the waveform over time. In these plots the resultant current measured from hundreds to thousands of consecutive voltammetric scans can be observed (Michael et al. 1998).

During the holding period, dopamine molecules can accumulate on the electrode surface. This process is driven by the negatively charged electrode surface, promoting adsorption of dopamine and other positively charged chemical species. These adsorbed dopamine molecules begin to become oxidized during the anodic voltage sweep to dopamine-o-quinone, specifically, once the potential exceeds +0.15 V versus Ag/AgCl, with the peak reaction occurring at approximately +0.7 V (see Fig. 5.1C). During the cathodic sweep, dopamine-o-quinone is reduced back to dopamine, with the peak reaction occurring at approximately –0.3 V (see Fig. 5.1C). It should be noted that using anodic limits (i.e., maximum voltage during the anodic sweep) greater than +1.0 V augments the sensitivity of dopamine detection by promoting the formation of oxygen-containing functional groups at the surface of the electrode, which provide a substrate for adsorption of negatively charged molecules (Hafizi et al. 1990, Heien et al. 2003, Keithley et al. 2011, Bath et al. 2000).

Application of the triangular voltage waveform generates not only faradaic current but also a very large charging or background current (~1,000 times the amplitude of the faradaic current elicited by physiological change in extracellular dopamine concentration). The background current, which is proportional to the scan rate, results from double-layer charging at the electrode surface (i.e., when the potential across the carbon fiber changes, charged molecules in the surrounding area become attracted to the fiber and coat it on either side). This effectively results in the formation of a capacitor across the electrode carbon fiber, which yields a charging (background) current. Fortunately, this current is relatively stable between scans, so the background from a baseline period can be subtracted to reveal changes in faradaic current (Bard and Faulkner 2001, Howell et al. 1986).

Factors that can interfere with measurement of dopamine oxidation currents include other electroactive molecules and pH changes at the electrode surface (Robinson et al. 2008). For example, the antioxidant ascorbic acid, which is found in large concentrations in the extracellular space (~0.5 mM) not only has an electrochemical oxidation/reduction signature similar to dopamine but also can reduce dopamine-o-quinone back to dopamine, resulting in a smaller reduction peak during the cathodic scan (Robinson et al. 2008, Sternson et al. 1973). Application of a high scan rate pushes the ascorbic acid CV away from the dopamine domain since ascorbic acid has slower electron-transfer kinetics, causing its CV to be more prolonged or broader than that of dopamine (Baur et al. 1988, Ewing et al. 1982). Local pH changes resulting from neuronal activation can also influence electrochemical signals since they shift the equilibrium of electrode surface reactions contributing to the background current, and the pH CV can overlap with the dopamine CV, obscuring the dopaminergic electrochemical signal (Venton et al. 2003). One way to remove this interference is to subtract the current obtained at a potential sensitive to pH but not dopamine from the current at the peak dopamine oxidation potential during the data analysis step (Michael et al. 1998, Venton et al. 2003). It is also possible to employ more sophisticated statistical approaches to separate analytes (Heien et al. 2005; see the "Chemometrics" section).

We focus on the basic theory and concepts underlying the electrochemical technique of FSCV with a primary focus on the neurotransmitter dopamine. It should be noted that while changes in norepinephrine and serotonin levels can also be detected using FSCV, obtaining chemical selectivity between these three monoamines is challenging. For most published studies to date, this problem has been circumvented by making measurements in brain regions where there is relative enrichment of the monoamine of interest (e.g., Hashemi et al. 2009, Heien et al. 2005, Park et al. 2011).

Instrumentation

Several considerations are required to ensure that the threshold of detection of the electrochemical instrumentation is adequate to measure subsecond physiological signals in the nanoampere range, including (1) low electrical noise, (2) sufficient signal amplification, and (3) high-resolution data collection capabilities:

1. To reduce noise that may overwhelm and/or distort the low current signal, voltammetric recordings should be conducted in a grounded Faraday cage to block external electrical fields. Depending on the experiment, this could mean having a conductive

metal mesh enclosure around an anesthetized recording setup, or a solid metal box large enough to contain an operant chamber and all the components required for the operant chamber and voltammetry recordings to function properly. Operant chambers can also have their major metal parts replaced with generic acrylic plastic. All the electrical components should be properly grounded even if located in a Faraday cage.

2. Small faradaic currents are quite vulnerable to distortion from surrounding electrical noise. To prevent this, the current detected by the working electrode should be both converted to voltage and amplified. This process should be carried out as close to the current source as possible to mitigate exposure to surrounding electrical noise.

3. Data acquisition hardware included in the system must have the capability to collect the continuous analog signal from the working electrode and convert it to a high-resolution digital representation that is then streamed as it is collected from the working electrode to the front-user interface and stored until subsequent analysis.

The basic instrumentation components for a contemporary voltammetry set up are diagrammed in Figure 5.2A. To control the system, a front-user interface equipped with a monitor is required. This interface, which governs all input/output parameters and has short-term data storage, can be a personal computer (PC) or a PC-based platform (e.g., PXI, National Instruments). Within the front-user interface is the data acquisition (DAQ) hardware, a device or card that functions as a digital/analog input and output module. It contains a digital-to-analog converter (DAC) and an analog-to-digital converter (ADC). The DAC is used to convert the digital waveform signal sent by the front-user interface to an analog signal that is ultimately transmitted to the working and reference electrodes. The ADC converts the analog input from the electrodes to a digital signal that is then stored in the front-user interface. The DAQ also includes precision timers that allow for the execution of complex waveforms and synchronization of data collection with (brain) stimulation protocols. The DAQ is physically integrated with a PC using a peripheral component interconnect (PCI) bus slot, which is located on a computer motherboard. The PCI bus is a hardware interface used to connect the DAQ to the computer; however, because the availability of such hardware is becoming more restrictive (including PCI express) in newer desktop and laptop computers, USB interfaces for DAQ are now common, as is the use of dedicated PC-based platform models.

The DAQ is connected to a connector interface (similar to a breakout box) that is usually externalized from the front-user interface. This connector interface serves as a physical shield that protects the inputs and outputs from electrical noise, organizes all the input/output cables, and provides ports for the power supply, headstage, commutator (electrical swivel), stimulator, and other equipment (e.g., behavioral program digital inputs that are sent to the front-user interface). A number of signal conditioning functions are also implemented at this stage, including filtration of the waveform timing signal and isolation of the digital and analog signals. From the connector interface, voltage from the power supply and the waveform signal from the DAQ go through the commutator to the headstage, which

(a)

(b)

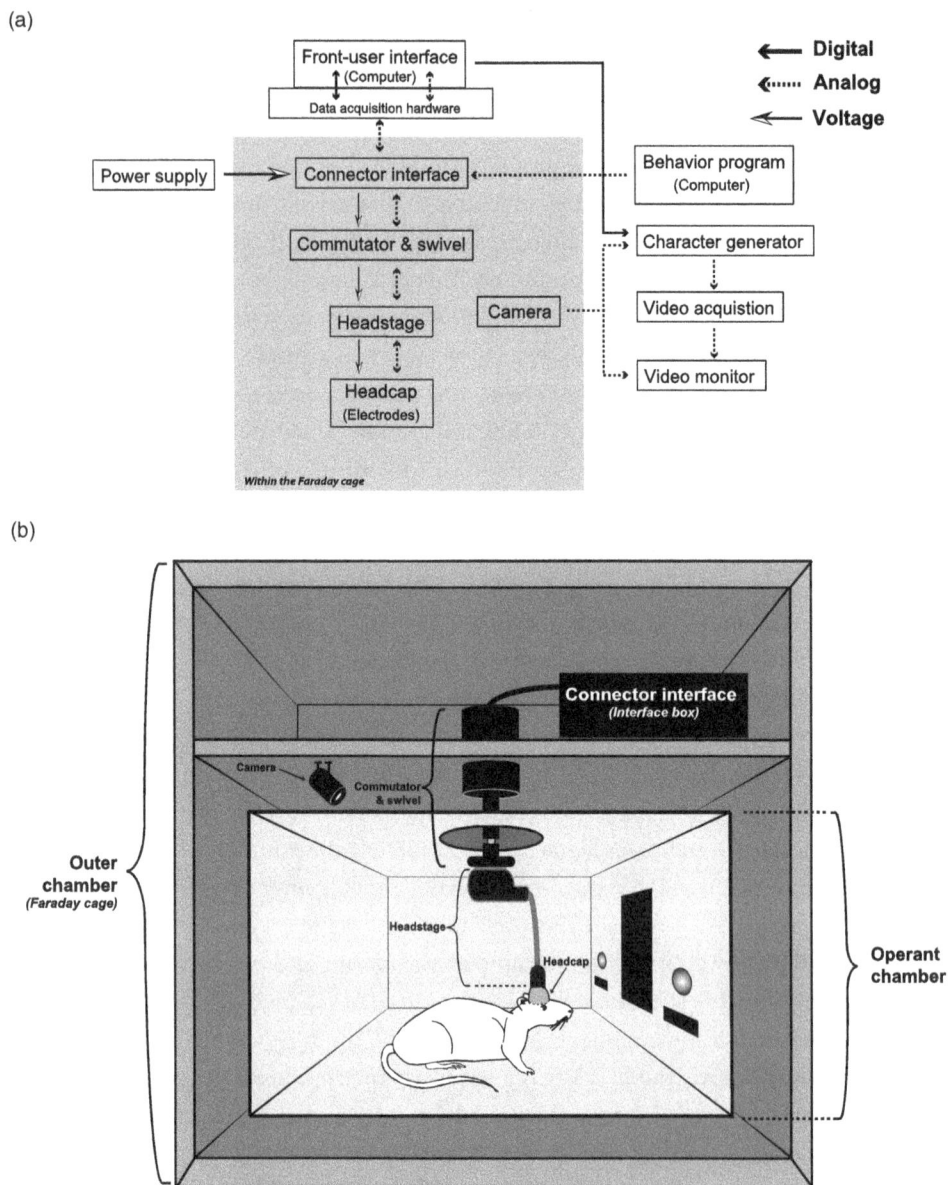

FIGURE 5.2 Instrumentation and FSCV setup. A: All major components of the instrumentation are depicted as a block diagram with their digital and/or analog connections. The gray box denotes the instrumentation that resides within the Faraday cage. B: The basic freely moving animal FSCV setup is illustrated excluding the accompanying front-user interfaces that control the voltammetry and behavior programs. Note that all the components in Figure 5.2B are also in Figure5.2A within the gray box. See accompanying detailed description in the text.

is connected to the electrodes. At the headstage, the current from the working electrode is converted to voltage by an operational amplifier with feedback resistor, providing sufficient gain to generate a signal in the low-volt range (up to ±10 V). This signal is low-pass-filtered to reduce any high-frequency electrical noise before it is sent to and stored by the front-user interface. In the case of a freely moving animal preparation, the headstage must be small, lightweight, and able to attach securely but easily to the animal via a head-mounted connector that is also wired to the implanted working and reference electrodes. It is also useful to have a video acquisition setup for post hoc analysis of behavior during the voltammetry recording as an ancillary to the behavioral data recorded by the behavior program. The video acquisition setup would include a camera positioned inside the operant chamber, which sends a live video feed to both the character generator and video monitor. The character generator receives the video feed and voltammetry recording information and imprints the voltammetry scan file number onto the video, which it then sends to the video acquisition hardware (e.g., a DVD recorder or DVR) and monitor. The result allows the experimenter to view (and record) the live feed of the experiment with the scan number imprinted in the video. Although timestamps for behavioral events are integrated into the voltammetry files, the video acquisition setup in conjunction with the character generator is an alternative (although less accurate) way to find timestamps of specific behavioral events. The illustration in Figure 5.2B is an example of a voltammetry system hardware setup that would be used to record dopamine signaling in a freely moving rat performing a task in an operant chamber contained within an outer chamber (Faraday cage) that serves as an electrical shield.

Electrodes

As animals learn during behavioral tasks it is likely that there will also be associated changes in dopaminergic signaling. To measure phasic changes not only in one behavioral session but across learning during potentially complex behavioral tasks, it is necessary to use a voltammetric sensor that is equipped to withstand the rigors of long-term experiments. Here we describe how to construct a chronically implantable microsensor (Fig. 5.3) that enables the monitoring of alterations in phasic dopamine signaling over the course of behavioral experiments that last for several months.

Chronically Implantable Carbon Fiber Microelectrodes. These electrodes (Clark et al. 2010) are lowered into the desired brain region and cemented in place on the rat's skull, allowing for repeated recordings across time. This approach permits sampling in the same location within a single animal throughout the course of complex behavioral experiments. Before beginning electrode construction, soak carbon fibers in 100% isopropanol. Use a razor blade to cut polyimide-coated fused silica (20 μm inner diameter, 90 μm outer diameter, 12 μm polyimide coating) to the desired length based on the intended target brain region (e.g., 12 mm for rat nucleus accumbens or 8 mm for rat dorsolateral striatum). The length needs to be sufficient for the depth that it will be lowered into the brain, plus the thickness of the skull, and some overlap with the silver pin in construction. Using a stereoscope, be sure that at least one end of the silica has a smooth flat surface on which to build the recording end of the electrode. We find that the best cutting technique to obtain a smooth end involves cutting the silica on the lab bench on top of two sheets of paper (marked with a line the length of the desired fragments) while angling the razor blade at ~45 degrees above the counter and rolling the silica toward you with the razor blade while pressing down. Soak silica fragments in 100% isopropanol just before beginning threading.

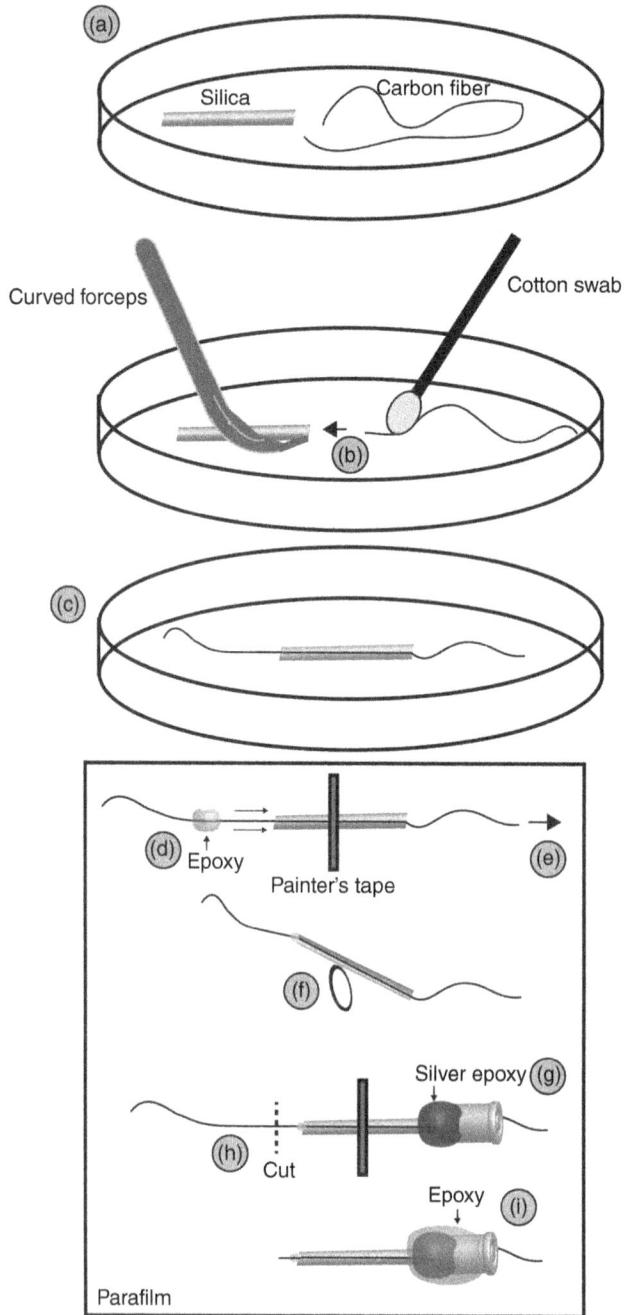

FIGURE 5.3 Chronic electrode construction. Setup (A), method (B), and final product (C) of threading the carbon fiber through the silica. D–F: Application of clear epoxy and positioning of the epoxy-sealed, threaded silica for drying. Silver epoxy placement (G), trimming of the carbon fiber (H), and the final clear epoxy sealing step (I). See accompanying detailed description in the text.

Prepare a "working dish" to thread the carbon fiber into the silica by filling a Petri dish about halfway (8 mm deep) with 100% isopropanol, and place one silica fragment and one carbon fiber (at least 8 cm long) into the isopropanol (see Fig. 5.3A). Working under the stereoscope, using a cotton swab to manipulate the carbon fiber and a pair of curved pointed forceps (FST 7) to hold the silica, push one end of the carbon fiber into the opening at the end of the silica fragment (see Fig. 5.3B). The distance from the end of the carbon fiber at which you press down with the cotton swab will change the height of the carbon fiber end above the bottom of the Petri dish; manipulating this can help in getting the fiber threaded into the opening at the end of silica that is slightly raised above the bottom of the Petri dish (silica is raised this way when it is positioned away from the center of the Petri dish, thus taking advantage of its convex shape). Once it is threaded all the way through the silica, leave the majority of the carbon fiber length extending beyond the end that you desire to be the recording end, and just about 1 cm extending beyond the other end of the silica (see Fig. 5.3C). Take the carbon fiber-threaded silica out of the isopropanol-filled dish and place it in another Petri dish to dry overnight before proceeding.

Using a very thin sliver of painter's tape, attach the center of the threaded silica to a piece of Parafilm (~ 3 cm × 3 cm). Mix a small amount of 5-minute clear epoxy (1:1 ratio of resin to hardener) using a toothpick until the mixture thickens slightly (~30–45 seconds), and then place a dab (1–2 mm in diameter) on the exposed carbon fiber 3–5 mm away from the desired recording end of the silica (see Fig. 5.3D). While observing under the stereoscope, pull the exposed carbon fiber from the nonrecording end of the silica and watch as beads of epoxy stick to the carbon fiber at the other end and travel into the silica (see Fig. 5.3E). Lift one end of the painter's tape and flip it over onto itself, making a loop that is still adhered to the Parafilm with the threaded silica positioned on top of the loop. Then continue pulling the carbon fiber until the last remaining beads of epoxy have been drawn into the silica (see Fig. 5.3F). Ideally this will result in about one third of the silica shaft being filled with epoxy, the end of the silica tipped with a rounded dome of epoxy, and very little epoxy on the shaft of the silica. It is acceptable to use electrodes that do not have this epoxy dome so long as the end of the silica shaft is completely sealed with epoxy and there are no jagged edges. Any epoxy that gets on the shaft of the electrode can be wiped off using a toothpick that already has dried epoxy on it (this prevents wood fibers from sticking to the electrode). Allow epoxy to dry while electrode tips remain propped up on the painter's tape loop above the Parafilm surface overnight. Keep electrodes in a drawer or closed container during all drying steps to prevent dust from sticking to the epoxy.

Unloop the painter's tape and reuse it to retape electrodes flat on a Parafilm surface in such a way as to avoid the dried epoxy bubble on the Parafilm. Mix a small amount of silver epoxy (equal parts resin and hardener) with a toothpick and place a dab (~3 mm in diameter) at the nonrecording end of the silica, halfway covering the end of the silica and halfway covering the carbon fiber. Place a female-only silver pin into the silver epoxy parallel to the silica, with the female end opening facing away from the silica. Add a bit more silver epoxy on top of the silver pin and smooth so that the closed end of the silver pin is completely

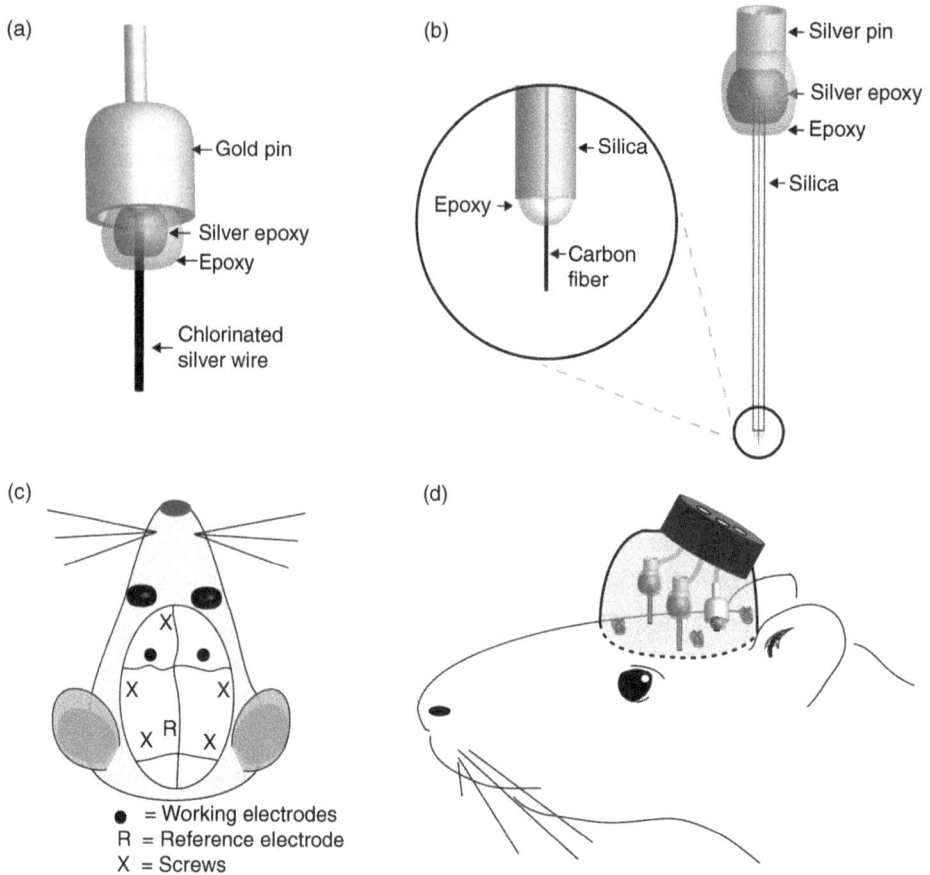

(a)

Gold pin

Silver epoxy
Epoxy

Chlorinated
silver wire

(b)

Silica

Epoxy →

Carbon
fiber

Silver pin

Silver epoxy

Epoxy

Silica

(c)

• = Working electrodes
R = Reference electrode
X = Screws

(d)

FIGURE 5.4 Electrodes, their surgical placement, and final positioning within the headcap. The chronically implantable carbon fiber microelectrode (A) and the silver/silver chloride reference electrode (B) are illustrated with their components labeled. C: A bird's-eye view of the exposed skull with suggested placements of the reference, working electrode(s), and anchor screws. D: All of the components are encased in dental cement to form a headcap with only the connector (Ginder) exposed on top.

coated in silver epoxy (see Fig. 5.3G). Allow the silver epoxy to dry overnight. This results in an electrical connection being formed between the silver pin and the carbon fiber.

Remove the tape from the silica. Using a stage micrometer under the microscope, trim the exposed carbon fiber at the recording end of the silica with a scalpel or spring scissors to yield an exposed length of fiber 150–200 μm (see Fig. 5.3H). Mix clear 5-minute epoxy (equal parts resin and hardener), place a dab (~3 mm in diameter) on the Parafilm, and place the silver epoxy/pin portion of the electrode directly on top of the epoxy. Take more epoxy and use it to cover the entirety of the silver epoxy (without getting much on the silica shaft; see Fig. 5.3I). Allow to dry overnight and then store on parafilm until use. Figure 5.4A shows a diagram of a constructed working electrode.

Reference Electrodes. Silver/silver chloride (Ag/AgCl) reference electrodes are a common type of reference electrode used in electrochemistry, particularly for biological applications. They are constructed by using silver epoxy to electrically connect a piece of silver wire to the female end of a gold pin that has both male and female ends. There should be ~5 mm of silver wire exposed outside of the gold pin/silver epoxy junction. Once this epoxy has dried (overnight), apply an outer coating of clear epoxy to cover the gold pin/silver epoxy junction, making sure not to get any on the length of exposed silver wire. Once this epoxy is dry, proceed to single-replacement reaction-mediated deposition of chloride ions on the surface of the silver wire by soaking the exposed portion of silver wire in a solution of sodium hypochlorite. To achieve this, fill a Petri dish with either sodium hypochlorite or chlorine bleach, and then put a piece of Parafilm over the top in such a way that the Parafilm is in direct contact with the surface of the solution. Poke the silver wire ends of the reference electrodes through the Parafilm and push them in up to the epoxy junction. Allow them to soak overnight, then remove and store by pushing the male end of the gold connector into a piece of Styrofoam. It is best to handle the electrode by the gold pin so as not to scratch any silver chloride from the silver wire surface. Electrodes can be dropped into isopropyl alcohol for ~30 seconds just prior to implantation to remove any debris that might have settled on the surface during storage. Figure 5.4B shows a diagram of a constructed reference electrode.

Animal Preparation

Although this chapter is focused on describing how to perform FSCV in freely moving rodents, the electrodes and FSCV parameters we describe can also be used in anesthetized animals and brain slice preparations. For anesthetized preparations, considerations for the type of anesthesia used must be made, as some anesthetics (e.g., isofluorane) are known to influence and possibly inhibit neuronal signaling.

All stereotaxic surgeries are carried out using aseptic techniques under anesthesia (ketamine/xylazine and isoflurane are most commonly used). Because of the variable requirements of individual animal use protocols and specific requirements at certain institutions we have omitted specific details of aseptic techniques, methods of anesthesia, and pre- and postoperative care.

Once the rat is in an appropriate plane of anesthesia, shave its head and secure in the stereotaxic frame. With scissors, cut a 1.5- to 2.0-cm-long slightly oval area on the scalp beginning just behind the eyes and ending at the ears. Using small scissors and a curette, scrape and cut away remaining connective tissue on the skull. Once the skull is cleaned, use the curette to score the exposed skull in several different directions to produce a rough surface to enhance cement adhesion later. From now on keep the skull as dry as possible by regularly wiping with cotton swabs, absorptive triangles, gauze, and so forth. Using the drill bit attached to a stereotaxic-mounted drill (e.g., a Dremel hand drill, equipped with a #1 size burr drill bit), ensure that the head is level (i.e., if the drill bit is zeroed out at the skull surface at bregma in the dorsal/ventral [DV] axis, the DV coordinate of the drill bit when placed at lambda should also read 0). Once the head is adjusted by moving the nosepiece up or down so that the skull is level, bring the drill bit to the skull surface at bregma and zero coordinates in anterior/posterior (AP) and medial/lateral (ML) directions. Move the drill bit to the AP and ML coordinates for selected working electrode sites and drill holes through the skull at these locations. If you will be implanting a stimulating electrode, its hole should also be drilled at this time. Using a manual stereotaxic-mounted twist drill with high-speed twist drill bits, make holes for at least three anchor screws (using four or five screws is recommended) in a configuration that will maximize the spacing between the screws and keep them at least a few millimeters away from the holes for the recording electrodes. Drill one additional hole for the reference electrode a few millimeters from one of the screw holes (see Fig. 5.4C).

Using forceps and a watchmaker's screwdriver, secure the screws in their holes, leaving at least 1 mm of threads visible between the head of the screw and the skull so that the cement will fill the space underneath the head of the screw. Flush out all remaining holes with sterile saline from a syringe fitted with a 26-gauge needle, and then dry the skull. Carefully clear the dura and make sure that there are no small bone fragments or pieces of tissue inside the holes, as working electrodes are susceptible to breakage during lowering if they run into bone debris or dura on the way into the brain. Using forceps, grip the reference electrode by the gold pin and place the silver wire end into its appropriate hole. The epoxy bubble should hit the skull and prevent it from going any farther into the brain. Dry up any blood or fluids that have accumulated on the skull surface, mix a small amount of acrylic dental cement so that it is moderately viscous and won't spread across the skull easily, and apply it to the base of the reference electrode where it meets the skull using a 1-mL syringe with a 20-gauge needle. At the same time, try to secure the reference electrode to the nearest screw with dental cement. It is very important during this stage to make sure that the cement does not travel across the skull into any of the other holes that have been drilled. A metal spatula can be used to manipulate the cement as it dries and push it away from any holes. Allow this cement to dry completely before once again flushing the remaining holes and drying the skull.

Attach a working electrode to a stereotaxic arm outfitted with a brass rod that has a male silver pin soldered to the end. Because of the variability in electrode construction, the brass rod will need to be bent or manipulated until the silica shaft of the electrode is completely vertical. Use vertical edges on the stereotaxic to check that the electrode is aligned with them in at least three planes. Align the electrode so that it is directly above the center of the working electrode hole. Plug the reference electrode into a headstage, and then connect the headstage to the brass bar holding the working electrode using an alligator clip. Open the waveform generator program and apply the desired waveform at 10 Hz to the electrodes. Turn on the power supply (at this point a background signal should not be present as the circuit is not complete). Dry any fluid in the hole and on the skull one final time, and then begin slowly lowering the electrode until a background waveform appears and reaches a maximum (the entire carbon fiber has made contact with the surface of the brain). At this point stop lowering and zero the DV coordinates. Proceed to lower the working electrode slowly until you are ~0.2 mm above the final desired coordinate. Mix a small amount of cement and apply it to the junction where the working electrode enters the skull, being sure to secure the epoxy/silver pin portion of the electrode body to the skull and not just the silica shaft. As previously, use viscous cement to avoid cement spreading across the skull. Once the cement is applied, lower the electrode to its final DV coordinate and allow the cement to completely dry before doing any further manipulations. If implanting two working electrodes, mount both of them on the stereotaxic at the same time and follow all subsequent steps in parallel so you can cement them both in at once. If you require a stimulating electrode, it should be implanted and secured with acrylic cement at this time. Once the cement is dry, turn off the power supply, stop the generation of the applied waveform to the recording electrodes in the waveform generator program, disconnect the working electrode(s) and then the reference electrode from the headstage.

Now the reference and working electrodes can be attached to their corresponding wires on the desired connector (e.g., DataMate or Ginder) that will be used to attach the animal to the headstage in the behavioral chamber. Mix up a large amount of cement and build a headcap that covers the electrodes and all the wires and the electrical connections between them and allows for the connector and stimulator pedestal (if applicable) to be secured at their bases, such that the connector is the only component exposed (see Fig. 5.4D). Even if you are not implanting a stimulating electrode, securing a stimulating-electrode pedestal or similar connector within the headcap provides an anchor point for the headstage leash, reducing torque if using a DataMate connector. This is not required if using Ginder connectors. Once the cement is dry, the animal can be removed from the stereotaxic and allowed to recover.

Other Experimental Applications Used in Conjunction with FSCV

FSCV can be combined with other techniques or applications in freely moving animal preparations. Examples of these combined approaches can be found in the following references (asterisks indicate an approach that uses chronically implanted electrodes as described in this chapter):

- Chemogenetics (Ferguson et al. 2011*)
- Intracranial self-stimulation (Garris et al. 1999, Kruk et al. 1998)
- Single-unit electrophysiological recording (Cheer et al. 2005)
- Microdialysis (Budygin et al. 2000)
- Site-specific microinjections of pharmacological agents (Wanat et al. 2013*)
- Ionophoretic delivery of pharmacological agents (Herr et al. 2012)

- Response to novel environments (Rebec et al. 1997)
- Pavlovian behaviors (Flagel et al. 2011*, Sunsay and Rebec 2008)
- Instrumental responding for food (Roitman et al. 2004, Wanat et al. 2010*)
- Intravenous drug self-administration (Phillips et al. 2003b, Willuhn et al. 2012*)
- Concurrent-choice decision making (Gan et al. 2010*)
- Aversive stimuli (Badrinarayan et al. 2012, McCutcheon et al. 2012, Oleson et al. 2012*)
- Maze-task learning (Howe et al. 2013*)

Data Collection
Background Currents and Cycling Electrodes

"Cycling" of the electrodes before data collection involves repetitive application of the voltammetry waveform that will be used during data collection to the working electrode in order to prime the electrode surface for recording. In addition to the capacitive charging current across the electrode surface, the redox current from electrolysis of the surface functional groups on the carbon fiber also influences the background current. Thus, cycling the electrode allows for the oxidation dynamics of surface functional groups to reach equilibrium. This yields the highly stable maximal amplitude background current that will be subtracted from recordings. We typically perform one session of electrode cycling on a day that precedes experimental testing in order to gauge the quality of implanted electrodes. Electrodes are then cycled before all behavioral sessions during which voltammetry data will be collected. To cycle electrodes, plug the connector on the animal's headcap into the appropriate end of the headstage, then plug the headstage into the commutator in the behavioral chamber to be used. Using the desired software to control voltammetry waveform generation (in our case, Tarheel CV), set up the program to apply the desired triangular waveform (as in Fig. 5.1A) at 10 Hz and subsequently turn on the power supply so that the waveform signal generated is applied to the electrode. Use the oscilloscope to observe the shape of the waveform of the background current. Typical background waveforms have a broad peak (point 1 in Fig. 5.5A) early in the anodic scan that has the largest amplitude and a secondary peak, truncated at the voltage switching potential (i.e., at +1.3 V; point 2 in Fig. 5.5A). The position, amplitude, and shape of these peaks provide indications of the quality of the implanted electrodes and whether we should expect to get viable recordings or not. In some cases, the position of the broad peak in the background waveform is delayed (i.e., the peak is shifted to the right). Sometimes this is observed at the onset of experimentation, and other times this shift can occur in the middle of an experiment. For example, in Figure 5.5A the background waveform observed has changed from one recording to the next, over the course of a few days. The background waveform taken from a session later in the experiment (late unshifted: *black dashed line*) is delayed compared to that taken from a session

(a)   Background CVs

(b)   Pellet response CVs

(c)   Pellet CV summary

Condition	Peak DA Voltage (V)	R²vs. DA standard
—— Early-Unshifted	+0.66 V	0.98*
- - - Late-Unshifted	+0.81 V	0.49
—— Late-Shifted	+0.66 V	0.96*

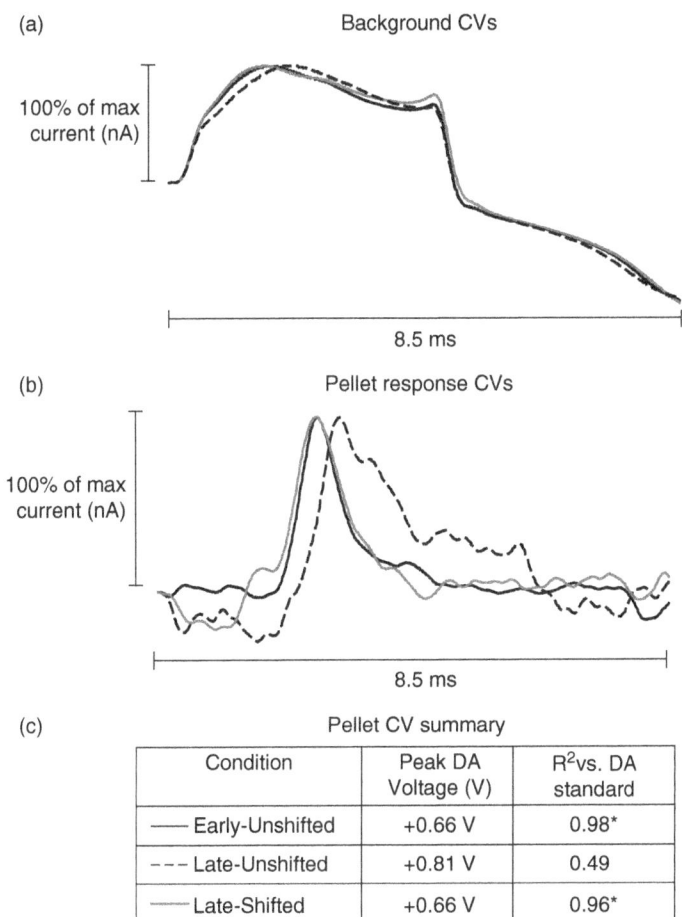

*CV was highly correlated (linear regression; $R^2 \geq 0.75$) to that of a dopamine (DA) standard

FIGURE 5.5 Effects of the reference holding potential reduction on the waveform shape. The background current (A), with its broad **(1)** and sharp **(2)** peaks indicated with arrows, and the faradaic current (B) measured in response to reward presentation from an individual rat both early and late in the time course of an instrumental behavior experiment when the electrode required shifting. C: Summary of the oxidation potential of dopamine measured during reward response at early and late time points in the experiment (these time points were days apart). In the early condition, no adjustments to the waveform voltage application were necessary (*black solid line*) since there was a high correlation to a dopamine standard. At a later time point the oxidation of dopamine occurred at a higher voltage and thus did not match the dopamine standard (*dashed black line*). This was corrected by shifting the applied potential by +0.200 V (*gray solid line*). Current (nanoamperes) was normalized to the maximum current observed in each CV for visual comparison between conditions.

earlier in the experiment (early unshifted: *black solid line*). This shift in the waveform is caused by a ~0.200-V reduction in the reference holding potential, possibly due to a loss of silver chloride over time as a result of interactions with the surrounding biological tissue (Clark et al. 2010, Heien et al. 2005, Moussy and Harrison 1994, Phillips et al. 2003a). Therefore, under these conditions the applied potential will be lower than anticipated, so the peak dopamine oxidation potential will be larger than usual (greater than +0.6 V as it is for late unshifted; Fig. 5.5B). Application of a +0.200-V DC offset to the waveform will compensate for the change in reference potential (compare late unshifted to late shifted; Fig. 5.5C) and normalize the background waveform. Once this compensatory shift has been made to the waveform (only if necessary) raise the frequency of waveform application to 60 Hz and allow equilibration of the background waveform. Using the oscilloscope, wait until the peak of the background waveform stops growing in amplitude (this takes 15–30 minutes depending on the number of times the animal has been previously cycled). At this point, switch the frequency of waveform application back to 10 Hz and allow for the same equilibration to occur in the opposite direction; the waveform will decrease in amplitude and then stabilize (this will require 40–60 minutes).

It is also important to assess whether offsetting the voltage by +0.2 V causes over-saturation. When this occurs, the peak current at the transition point in the voltammetric waveform (+1.3 V top of the triangle) will increase, resulting in the elevation of the sharp peak of the background waveform (location 2 in Fig. 5.5A) relative to the broad peak (location 1 in Fig. 5.5A). If the sharp peak is at the same height as or higher than the broad peak, this is indicative of the waveform being "overshifted." In this case, allow the electrode to cycle longer to ensure that this effect persists before reverting back to a 0-V voltage shift. In the most dramatic cases, the background waveform will appear "cut off" and will not be visualized in its entirety on the oscilloscope. When this happens it may not be possible to shift that electrode. If an electrode remains "overshifted" during a recording, the peak dopamine oxidation will occur at a more negative potential (i.e., less than +0.6 V) and these signals may not be detected by the statistical methods used to analyze the data.

Recording Voltammetry Data During Behavior

After electrodes have been cycled, we recommend taking a few test recordings to double-check that your background current remains stable over time as well as to ensure that all settings in the voltammetry program are selected correctly. To ensure that you can detect changes in dopamine signaling, you may record phasic dopamine release elicited via electrical stimulation of dopamine afferents or an external event such as presentation of an unexpected reward. Verification that the signals observed are, in fact, due to released dopamine can be carried out quickly by measuring the extent of correlation of the CV obtained in response to the stimulus presentation to that of the CV of a dopamine standard of known concentration ($r^2 \geq 0.75$ by linear regression). This standard is a part of a training set that would be created in house by measuring electrically stimulated dopamine release in an awake rat (see the "Data Analysis" section). Once you are content with the quality of your test recordings and positive control recordings, you may proceed to record during testing.

Once ready to start a behavioral session, be sure that data collection software is set to make multiple file collections such that recordings will be made throughout the duration of your session. If recording video of the session, begin recording before starting any of the voltammetry or behavior. Start the voltammetry recording and then start the behavioral program (e.g., Med Associates, MED PC software). Voltammetry data will now be collected throughout the session. Behavioral digital input/output signals should be registered and time-stamped within your collected data files so that you can later go back and look at signals associated with specific behavioral measures, cue presentations, and so forth. Upon completion of the behavior session and data collection, turn off the power supply to the electrodes and then stop waveform application and close the software. Only after these steps have been completed may the animal be disconnected.

Data Analysis

Histology

Once the experiment is complete, to verify that the electrodes are indeed in the targeted area, histology must be performed on the brains to confirm electrode placement. To make the carbon fiber location visible (fiber is too small to make visible marks in the tissue on its own), once the rat has been deeply anesthetized, and immediately prior to transcardial perfusion with saline and 4% paraformaldehyde (PFA), create an electrolytic lesion by passing a current (~70 nA) through the carbon fibers for 20 seconds. After the perfusion, remove the brain from the skull and store in 4% PFA. Prior to slicing, sink the brain in 15% sucrose in 1× phosphate-buffered saline (PBS), then in 30% sucrose/1× PBS. Section brains (30–60 μm), mount slices onto glass slides, and stain in a 4% cresyl violet solution. Locations of the electrolytic lesions from the carbon fiber electrodes can then be determined using a light microscope and a brain atlas (e.g., the rat brain atlas by Paxinos and Watson, 5th ed.).

Voltammetry Data Processing

Phasic responses evoked by any behavioral measure or cue/operant presentation that is associated with digital behavioral input/output signals can be studied. Alterations in phasic dopamine transients during different time windows can also be investigated. In our lab we typically study phasic responses associated with specific behaviors or stimuli (e.g., reward delivery, reward-predictive stimulus). This type of analysis is described below.

Cutting Data Files. First, files must be generated in which voltammetry data spanning the response of interest have been "cut out" of the raw data files and aligned so that the response of interest occurs at the same time point in all files. We use an in-house program (written using LabVIEW) that strings all recorded files from a single session together and produces peri-event data "snips" of the desired time window around the digital signal specified. This process is repeated for each of the signals of interest.

Chemometrics. Chemometrics is a technique used to extract signals associated with phasic dopamine release events from data. This technique employs a data training set that includes responses to multiple dopamine and pH concentrations used as standards.

Principal component analysis is then used to construct a mathematical model, which estimates dopamine concentrations from experimental data sets (see Keithley et al. 2009 for review). Aspects of the signal that cannot be attributed to dopamine such as pH (or other chemicals if they are included in the training set) are included in a residual. If this residual includes deterministic information, then the model is rejected as it is not sufficient to account for the data. In practice, such a scenario is rare, especially when the peri-event data "snips" are relatively short (<90 s; Heien et al. 2005).

Plotting and Statistical Analysis. Prior to chemometric analysis, we apply a low-pass filter (fourth order, 2-kHz Butterworth filter) to remove high-frequency noise from our signals, and then we typically smooth data obtained by chemometric approaches using a five-point running average before plotting or performing any statistical analysis. Once smoothed, common measures that can be studied are the signal peak height, area under the curve, time to peak, half-life of signal decay, or frequency of spontaneous signaling events (using criteria that define what makes up a spontaneous event). Individual trials can be studied, as well as average signals obtained after specific behavioral events or stimulus presentations. These events can be analyzed using standard statistical procedures and plotted with any of the commonly used graphing and statistical software packages available (e.g., Excel, Graphpad Prism, Matlab, SPSS).

Frequently Asked Questions

Q: Where can all the necessary instrumentation hardware and software be purchased?

A: To date there is no general FSCV hardware vendor. Laboratories with existing FSCV setups have relied on in-house machine shops to obtain the necessary equipment. Software is also generally developed in house and can be written using graphical programming software such as LabVIEW (National Instruments).

Q: What is causing our system to have electrical noise?

A: Unfortunately, there are no simple answers relating to a general issue of electrical noise, since its presence can be due to numerous sources. The best advice is to systematically check all the FSCV hardware to make sure that every step possible has been taken to electrically ground or shield the hardware and system as a whole.

Q: What happens if the carbon fiber is not within the length range specified (150–200 μm)?

A: The length of the carbon fiber determines its surface area, which is directly related to the number of molecules that can potentially adsorb to the fiber. Thus, the carbon fiber must be cut with these length limitations in mind, as there is a finite range within which the FSCV system can accurately measure faradaic current generated by redox reactions at the surface of the electrode. Furthermore, once the electrode is implanted, adjustments to the length of the carbon fiber cannot be made to alter the sensitivity of the electrode. A fiber length <150 μm will result in an insufficient surface area, and the electrode will not be sensitive enough to detect nanoampere changes in current. A carbon fiber >200 μm, on the other hand, may yield too much adsorption at its surface, causing oversaturation of the faradaic current beyond the detection capability of the system. One can speculate whether the fiber length is too short or long by the shape and amplitude of the background waveform viewed on the oscilloscope. The background waveform from a short fiber will look "blunted," whereas a waveform from a fiber that is too long will not fit on the oscilloscope screen and parts of it will be "cut off." Since it is difficult to describe in words, you can visualize this first hand by connecting electrodes of varying carbon fiber lengths (e.g., 100 μm, 150 μm, and 250 μm) to a voltammetry system one at a time and observing the shapes of each of the background waveforms when the electrode is submerged in saline (not water).

Q. How long after surgery should FSCV recording start?

A. In our experience (Clark et al. 2010), phasic dopamine signals in response to salient events (e.g., reward-predictive cues, unexpected reward delivery) begin to be detectable one month after electrode implantation. As a suggestion to save time, after postoperative recovery, you can start to behaviorally train the rat at a time point such that the first FSCV recording day falls on or right after the one-month-postsurgery mark. Also, as mentioned in the "Data Collection" section, be sure to include days to cycle the electrodes prior to the first recording day.

Q. How do I know when I should offset the applied voltage +0.2 V (i.e., shift the electrode)?

A. The topic of electrode shifting is covered in the "Background Currents and Cycling Electrodes" section of this chapter. An important point to reiterate from that section is that the electrodes must be cycled at a time prior to the first testing day in order to visualize the shape of the background waveform and prime the electrode surface for recording. It is the shape of the background waveform and the voltage at which peak oxidation occurs in a putative dopamine CV that determine whether the electrodes should be shifted (see Fig. 5.5).

Q. How do I test to see whether an electrode is measuring changes in dopamine concentration?

A. This step is covered in the "Data Collection" section, but briefly, testing for detection of phasic dopamine changes in an awake, moving animal requires exposing that animal to a familiar stimulus that has been historically reliable at eliciting phasic dopamine signaling. This includes, but is not limited to, presentation of reward-predictive cues, unexpected reward delivery, electrical stimulation of dopamine afferents, or intravenous injection of a pharmacological agent, such as cocaine, that causes an immediate increase in dopamine concentration. Signals generated from these types of stimuli are then to be compared to a dopamine standard to determine whether the signal is comparable to the electrochemical signature of dopamine.

Q. Why is there inconsistency in the electrode's signal or detection properties between test days (assuming instrumentation is functioning properly)?

A. Some days it takes longer for the electrodes' signal to stabilize (stop drifting during the cycling phase). Kinks in the headstage cable can disrupt the signal flow through the headstage, resulting in abrupt fluctuations of the applied voltage to the electrode. This sudden voltage fluctuation can cause it to take longer for the signal to stabilize. These kinks often result from the animal moving in the chamber and causing the cable to become coiled, or from the animal chewing on the headstage cable. Also, shifting the voltage offset after the electrode has already been cycling for a period of time will require extended cycling as a change in the voltage application causes drifting.

Q. The electrode signal is noisy today, but on previous days noise was not an issue.

A. This may be due to an incomplete connection between the headcap and the headstage, which causes fluctuations in the current, particularly when the animal moves and disrupts the connection even more. Shut the system down, unplug the animal, and reattach the headstage (you might also try a different headstage, if it's available). The noise could also be due to the presence of dust or debris in the connectors. Clean the gold pins on the headstage and in the connector on the top of the headcap, where the headstage connects, with 70% isopropyl alcohol and air-dry for 5–10 minutes before reattaching the animal to the system.

Q. The background waveform is oversaturated, but on previous recording days, oversaturation was not an issue.

A. Offsetting the voltage potential applied to the electrode can cause oversaturation of the waveform. If this is the first time the electrodes have been shifted, let it cycle for ~30 minutes at 60 Hz to determine whether the waveform will eventually settle. If it doesn't, shifting the electrode may not be possible. It is worth proceeding with the cycling process and attempting to confirm the electrode's ability to detect dopamine prior to the behavioral session. Another reason could be that there is water within the connection (e.g., from the water bottle dripping onto headcap components while the animal is in its home cage). Unplug the animal from the headstage and make sure that there is no moisture on the headstage connectors or on the headcap connectors. To be certain, clean both connectors with 70% isopropyl alcohol, air-dry for 5–10 minutes, and reconnect the animal to the system. If that doesn't change the waveform shape, try connecting the animal with a different headstage.

Q. Dopamine has been readily detected previously, but on some occasions there doesn't appear to be a phasic signal.

A. One possibility is that there was an incomplete connection between the headstage and electrodes via the headcap during the instances when a signal wasn't detectable. The only way to know whether this is the case is to reconnect the animal to the same or a different headstage and make sure the connections are complete. In the circumstance where phasic dopamine signals were readily detectable but declined in strength after multiple recordings, the integrity of the carbon fiber may be irreversibly compromised as a consequence of extended duration of exposure to voltage application.

Q. Why can't I detect dopamine?

A. Unfortunately, because FSCV is not a "plug-and-play" system, there are numerous components, factors, and assumptions involved with detecting nanoampere changes in current in an awake animal that is moving and performing a behavioral task. This limits the level of control the experimenter has on the system, and it follows that there are numerous reasons for not being able to identify phasic changes in dopamine concentration. The list includes (but is not limited to) issues with instrumentation (hardware and/or software), electrode construction, proper electrode placement, how long after surgery it has been, and the state of the animal that is being tested (e.g., experience with the behavioral task and/or motivational state). Therefore, care must be taken at every step, such as the construction of the FSCV system and electrodes, proper electrode placement into a dopamine-rich area in the brain, appropriately training and/or food restricting the animal, ensuring that cues and/or other stimuli and rewards associated with the experiment are salient and have meaning to the animal, and acclimating the animal to being tethered within the behavioral chamber. Therefore, similar to the issue of electrical noise (which can be a reason for not detecting phasic dopamine), a systematic approach of testing the equipment and experimental design is required to identify the issue(s) responsible for the lack of dopamine detection.

Q. How do I know if the electrode is located at its intended anatomical target?

A. The appropriate way (which is most certainly required for publication) to determine whether the carbon fiber was placed in the appropriate brain region or area is through histological confirmation (see the "Data Analysis" section).

Q. How do I use chemometrics to analyze my data?

A. Chemometrics is a complex data processing technique that requires further reading (*Chemometric Techniques for Quantitative Analysis* by Richard Kramer, and Keithley et al. 2011) to develop chemometric analyses appropriate for your needs. Briefly, to apply chemometrics to your voltammetric data, a training set of known (or reference) dopamine concentrations needs to be generated in your lab, using your equipment to account for the noise component of your FSCV system. For our purposes, since the voltammetric data we generate is from an awake, moving rat implanted with chronic microsensors, we created a training set of known dopamine concentrations by recording FSCV while electrically stimulating dopamine release in a rat freely moving in one of our operant chambers. This training set is then used to perform principal component analysis and regression (we used MATLAB software) to compare the reference dopamine concentrations to the neurochemical information obtained from a FSCV recording during an experimental behavioral session (Table 5.1).

TABLE 5.1 Necessary Equipment

Equipment	Potential Sources
INSTRUMENTATION	
Character generator	Decade Engineering (XBOB4)
Commutator	Crist Instruments (4-TBC-9S)
Data acquisition cards	National Instruments (PCI-6052E. PCI-6711)
ELECTRODES	
Carbon fiber	Goodfellow (Grade: 34-700)
Clear epoxy (5-minute)	Amazon.com (Devcon 20845)
Curved, pointed forceps	Fine Science Tools (11274-20/Dumont #7)
Gold pin	Newark (82K7794)
Polyimide-coated silica tubing	Polymicro (2000381-10M/ID 20 μm, OD 90 μm, coating 12 μm)
Silver epoxy	Allied Electronics (661-3536)
Silver pin	Newark (23K7802)
Silver wire	Sigma Aldrich (265586-10G)
SURGERY	
Anchor screws	McCaster-Carr (91793A052/0-80 thread, 1/8-inch-length slotted machine screw)
Data mate connector (male—implant)	Mouser.com or Newark.com (M80-8630642/ six-pin male with friction latch)
Data mate connector (female—headstage and dummy connector)	Mouser.com or Newark.com (M80-6810605)
Dental cement	Lang Dental (B1323/Ortho-jet BCA package)
Drill bits (electric hand drill)	Roboz Surgical Equipment (rs-6280c-1/size #1 bur)
Drill bits (hand twist drill)	McCaster-Carr (30585A72/high-speed twist drill bit, gauge 56)
Ginder connector	Ginder Scientific GS09 series nylon and cap nut

Further Reading

Bard, A. J., & Faulkner, L. R. (2001). *Electrochemical methods: Fundamentals and applications.* New York: John Wiley & Sons, Inc.

Kuhn, C. M., & Koob, G. F. (Eds.). (2010). *Advances in the neuroscience of addiction.* Boca Raton, FL: CRC Press.

Michael, A. C., & Borland, L. M. (Eds.). (2007). *Electrochemical methods for neuroscience.* Boca Raton, FL: CRC Press.

Wightman, R. M. (2006). Probing cellular chemistry in biological systems with microelectrodes. *Science* 311: 1570–1574.

References

Aston-Jones, G.,& Cohen, J. D. (2005). Adaptive gain and the role of the locus coeruleus-norepinepherine system in optimal performance. *J Comp Neurol* 493(1): 99–110.

Badrinarayan, A., Wescott, S. A., Vander Weele, C. M., et al. (2012). Aversive stimuli differentially modulate real-time dopamine transmission dynamics within the nucleus accumbens core and shell. *J Neurosci* 32(45): 15779–15790.

Bath, B. D., Michael, D. J., Trafton, B. J., et al. (2000). Subsecond adsorption and desorption of dopamine at carbon-fiber microelectrodes. *Anal Chemistry* 72(24): 5994–6002.

Baur, J. E., Kristensen, E. W., May, L. J., et al. (1988). Fast-scan voltammetry of biogenic amines. *Anal Chemistry* 60(13): 1268–1272.

Brazell, M. P., Kasser, R. J., Renner, K. J., et al. (1987). Electrocoating carbon fiber electrodes with Nafion improves selectivity for electroactive neurotransmitters. *J Neurosci Methods* 22(2): 167–172.

Budygin, E. A., Kilpatrick, M. R., Gainetdinov, R. R., & Wightman, R. M. (2000). Correlations between behavior and extracellular dopamine levels in rat striatum: comparison of microdialysis and fast-scan cyclic voltammetry. *Neurosci Lett* 281: 9–12.

Cheer, J. F., Heien, M. L. A. V., Garris, P. A., et al. (2005). Simultaneous dopamine and single-unit recordings reveal accumbens GABAergic responses: Implications for intracranial self-stimulation. *Proc Natl Acad Sci USA* 102(52): 19150–19155.

Clark, J. J., Sandberg, S. G., Wanat, M. J., et al. (2010). Chronic microsensors for longitudinal, subsecond dopamine detection in behaving animals. *Nature Methods* 7(2): 126–132.

Ewing, A. G., Wightman, R. M., & Dayton, M. A. (1982). In vivo voltammetry with electrodes that discriminate between dopamine and ascorbate. *Brain Res* 249(2): 361–370.

Flagel, S. B., Clark, J. J., Robinson, T. E., et al. (2011). A selective role for dopamine in stimulus-reward learning. *Nature* 469(7328): 53–57.

Ferguson, S. M., Eskenazi, D., Ishikawa, M., et al. (2011). Transient neuronal inhibition reveals opposing roles of indirect and direct pathways in sensitization. *Nat Neurosci* 14(1): 22–24.

Gan, J. O., Walton, M. E., & Phillips, P. E. M. (2010). Dissociable cost and benefit encoding of future rewards by mesolimbic dopamine. *Nat Neurosci* 13(1): 25–27.

Garris, P. A., Kilpatrick, M., Bunin, M. A., et al. (1999). Dissociation of dopamine release in the nucleus accumbens from intracranial self-stimulation. *Nature* 398(6722): 67–69.

Gerhardt, G. A. G., Oke, A. F. A., Nagy, G. G., et al. (1984). Nafion-coated electrodes with high selectivity for CNS electrochemistry. *Brain Res* 290(2): 390–395.

Hafizi, S., Kruk, Z. L., & Stamford, J. A. (1990). Fast cyclic voltammetry: improved sensitivity to dopamine with extended oxidation scan limits. *J Neurosci Meth* 33(1): 41–49.

Hashemi, P., Dankoski, E. C., Petrovic, J., et al. (2009). Voltammetric detection of 5-hydroxytryptamine release in the rat brain. *Anal Chem* 81: 9462–9471.

Heien, M. L. A. V., Khan, A. S., Ariansen, J. L., et al. (2005). Real-time measurement of dopamine fluctuations after cocaine in the brain of behaving rats. *Proc Natl Acad Sci USA* 102(29): 10023–10028.

Heien, M. L. A. V., Phillips, P. E. M., Stuber, G. D., et al. (2003). Overoxidation of carbon-fiber microelectrodes enhances dopamine adsorption and increases sensitivity. *The Analyst* 128(12): 1413–1419.

Herr, N. R., Park, J., McElligott, Z. A., et al. (2012). *In vivo* voltammetry monitoring of electrically evoked extracellular norepinephrine in subregions of the bed nucleus of the stria terminalis. *J Neurophysiol* 107: 1731–1737.

Howe, M. W., Tierney, P. L., Sandberg, S. G., et al. (2013). Prolonged dopamine signaling in striatum signals proximity and value of distant rewards. *Nature* 500(7464): 575–579.

Howell, J. O., Kuhr, W. G., Ensman, R. E., & Wightman, R. M. (1986). Background subtraction for rapid scan voltammetry. *J Electroanal Chem* 209(1): 77–90.

Jacobs, B. L., & Fornal, C. A. (1991). Activity of brain serotonergic neurons in the behaving animal. *Pharmacol Rev* 43(4): 563–578.

Jennings, K. A. (2013). A comparison of the subsecond dynamics of neurotransmission of dopamine and serotonin. *ACS Chem Neurosci* 4(5): 704–714.

Keithley, R. B., Heien, M. L., & Wightman, R. M. (2009). Multivariate concentration determination using principal component regression with residual analysis. *Trends Anal Chem* 28(9): 1127–1136.

Keithley, R. B., Takmakov, P., Bucher, E. S., et al. (2011). Higher-sensitivity dopamine measurements with faster-scan cyclic voltammetry. *Anal Chem* 83(9): 3563–3571.

Kramer, R. (1998). *Chemometric techniques for quantitative analysis*. Boca Raton, FL: Taylor and Francis Group.

Kruk, Z. L., Cheeta, S., Milla, J., et al. (1998). Real time measurement of stimulated dopamine release in the conscious rat using fast cyclic voltammetry: dopamine release is not observed during intracranial self-stimulation. *J Neurosci Meth* 79(1): 9–19.

McCutcheon, J. E., Ebner, S. R., Loriaux, A. L., & Roitman, M. F. (2012). Encoding of aversion by dopamine and the nucleus accumbens. *Front Neurosci* 6: 137.

Michael, D., Travis, E. R., & Wightman, R. M. (1998). Color images for fast-scan CV measurements in biological systems. *Anal Chem* 70(17): 586A–592A.

Montague, P. R., Dayan, P., & Sejnowski, T. J. (1996). A framework for mesencephalic dopamine systems based on predictive Hebbian learning. *J Neurosci* 16(5): 1936–1947.

Moussy, F., & Harrison, D. J. (1994). Prevention of the rapid degradation of subcutaneously implanted Ag/AgCl reference electrodes using polymer coatings. *Anal Chem* 66(5): 674–679.

Niv, Y. (2007) Cost, benefit, tonic, phasic: What do response rates tell us about dopamine and motivation? Ann NY Acad Sci 1004: 357–376.

Oleson, E. B., Gentry, R. N., Chioma, V. C., & Cheer, J. F. (2012) Subsecond dopamine release in the nucleus accumbens predicts conditioned punishment and its successful avoidance. *J Neurosci* 32(42): 14804–14808.

Palij, P., & Stamford, J. A. (1992). Real-time monitoring of endogenous noradrenaline release in rat brain slices using fast cyclic voltammetry: 1. Characterization of evoked noradrenaline efflux and uptake from nerve terminals in the bed nucleus of stria terminalis, pars ventralis. *Brain Res* 587(1): 137–146.

Park, J., Takmanov, P., & Wightman, R. M. (2011). *In vivo* comparison of norepinephrine and dopamine release in rat brain by simultaneous measurements with fast-scan cyclic voltammetry. *J Neurochem* 119: 932–944.

Paxinos, G., & Watson, C. (2004). *The Rat Brain in Stereotaxic Coordinates*. Elsevier-Academic Press.

Phillips, P. E. M., Robinson, D. L., Stuber, G. D., et al. (2003a). Real-time measurements of phasic changes in extracellular dopamine concentration in freely moving rats by fast-scan cyclic voltammetry. *Meth Mol Med* 79: 443–464.

Phillips, P. E. M., Stuber, G. D., Heien, M. L. A. V., et al. (2003b). Subsecond dopamine release promotes cocaine seeking. *Nature* 422(6932): 614–618.

Phillips, P. E. M., & Wightman, R. M. (2003). Critical guidelines for validation of the selectivity of *in-vivo* chemical microsensors. *Trends Anal Chem* 22(9): 509–514.

Pihel, K., Travis, E. R., Borges, R., & Wightman, R. M. (1996) Exocytotic release from individual granules exhibits similar properties at mast and chromaffin cells. *Biophys J* 71(3): 1633–1640.

Rebec, G. V., Christensen, J. R., Guerra, C., & Bardo, M. T. (1997). Regional and temporal differences in real-time dopamine efflux in the nucleus accumbens during free-choice novelty. *Brain Res* 776(1-2): 61–67.

Robinson, D. L., Hermans, A., Seipel, A. T., & Wightman, R. M. (2008). Monitoring rapid chemical communication in the brain. *Chem Rev* 108(7): 2554–2584.

Roitman, M. F., Stuber, G. D., Phillips, P. E. M., et al. (2004). Dopamine operates as a subsecond modulator of food seeking. *J Neurosci* 24(6): 1265–1271.

Sternson, A. W., McCreery, R., Feinberg, B., & Adams, R. N. (1973). Electrochemical studies of adrenergic neurotransmitters and related compounds. *J Electroanal Chem* 46(2): 313–321.

Sunsay, C., & Rebec, G. V. (2008). Real-time dopamine efflux in the nucleus accumbens core during Pavlovian conditioning. *Behav Neurosci* 122(2): 358–367.

Swamy, B. E., & Venton, B. J. (2007). Subsecond detection of physiological adenosine concentrations using fast-scan cyclic voltammetry. *Anal Chem* 79(2): 744–750.

Venton, B. J., Michael, D. J., & Wightman, R. M. (2003). Correlation of local changes in extracellular oxygen and pH that accompany dopaminergic terminal activity in the rat caudate-putamen. *J Neurochem* 84(2): 373–381.

Wanat, M. J., Bonci, A., & Phillips, P. E. M. (2013). CRF acts in the midbrain to attenuate accumbens dopamine release to rewards but not to their predictors. *Nature Neurosci* 16(4): 383–385.

Wanat, M. J., Kuhnen, C. M., & Phillips, P. E. M. (2010). Delays conferred by escalating costs modulate dopamine release to rewards but not their predictors. *J Neurosci* 30(36): 12020–12027.

Willuhn, I., Burgeno, L. M., Everitt, B. J., & Phillips, P. E. M. (2012). Hierarchical recruitment of phasic dopamine signaling in the striatum during the progression of cocaine use. *Proc Natl Acad Sci USA* 109(50): 20703–20708.

Electroencephalogram Recording in Humans

Julia W. Y. Kam and Todd C. Handy

Introduction

An electroencephalogram (EEG) is a scalp-recorded time series of data that captures population-level electrical activity that systematically varies with the internal state of the subject. Activity at each electrode in the recording montage is plotted in units of scalp voltage (in microvolts) over time (in milliseconds). In cognitive neuroscience, the most common use of EEG is to derive event-related potentials (ERPs), which are signal-averaged epochs of EEG time-locked to the onset of condition-specific stimuli or motor events. On the assumption that the brain's response to the event of interest remains relatively stable for the duration of a recording session, the signal averaging procedure preserves that response in the derived ERP while treating the event-uncorrelated activity in each EEG epoch as noise that is theoretically attenuated in direct proportion to the number of events included in the signal averaging (e.g., Perry 1966). Recently, however, there has been growing interest in studying "resting-state" EEG, recorded while an individual is sitting quietly, for clues to the individual's neurocognitive status when unengaged by an external task. This chapter speaks to issues germane to both ERPs and resting-state EEG.

As a methodology to apply in the laboratory or clinical setting, EEG-based techniques have a number of key strengths. First, and most commonly discussed, EEG provides exquisite temporal resolution with respect to the nature of ongoing activity related to neurocognitive functions, on the order of milliseconds. This allows the researcher or clinician to dissociate subtle but meaningful differences in the timing of synchronized neural processes between conditions and/or populations of interest. Less appreciated but no less useful is the fact that EEG signals also have a high degree of resolution in the amplitude (or voltage) domain, giving them an ability to reveal condition- or group-dependent differences in the magnitude of synchronized neural processes that is on par with EEG's temporal resolution. Finally, when contrasted with the costs, infrastructure, and human resource requirements of other functional brain imaging methods such as fMRI, PET, and MEG, an EEG/ERP recording system is comparatively inexpensive to purchase and can be easily managed by the individual user. This gives EEG-based methods a degree of accessibility simply not possible with fMRI and related techniques, an issue that is particularly relevant when considering the development of functional biomarkers for use in clinical diagnostics.

Critically, it is important to appreciate at the outset what EEG-based methods *cannot* reveal. Most notably, compared to fMRI, PET, and MEG, EEG and ERP signals provide only the coarsest of spatial resolution with respect to the signals involved. To make definitive anatomical statements about the cortical source of an EEG-based signal typically requires extensive converging evidence from other imaging methods (e.g., Heinze et al. 1994). Second, when recording EEG from the scalp surface, only population-level electrocortical activity is actually measured, and at that, only neuron populations that can generate far-field potentials, or potentials that can propagate the required distance from the cortical signal source to the scalp surface (as described later). Thus, EEG-based techniques remain blind to much in the way of meaningful cognitive-related electrocortical activity. Nevertheless, as long as these limitations are acknowledged, EEG-based methods are a reliable and powerful tool for researchers and clinicians alike. Indeed, despite the powerful new imaging technologies that have been developed over the past several decades, EEG promises to remain an indispensable technique in cognitive neuroscience and related fields.

Basic Theory

The EEG is the recording of spontaneous electrical activity from ensembles of neurons in the brain using macroelectrodes placed across the surface of the scalp. This stands in contrast to extracellular or intracellular recordings using microelectrodes, which measure the electrical activity of a single neuron or a few local neurons. While it is crucial to elucidate the workings of a neuron at an individual level, the simultaneous electrical responses of multiple neurons are equally important to understand as population activity is necessary to bring about any cognitive and behavioral processes. EEG provides a noninvasive technique to measure the summed activity of a large group of neurons. This section first reviews the cellular mechanisms underlying EEG, and then it describes two types of data generated by EEG recording.

Cellular Mechanisms

The electrical responses recorded noninvasively from the surface of the scalp reflect the summed synaptic potentials of many activated neurons. Specifically, EEGs are thought to record activity of primarily cortical pyramidal cells (Martin 1991), which are the main projection neurons in the cerebral cortex, with axons projecting both locally and to other regions of the brain. They receive mainly excitatory synaptic inputs, using glutamate as the neurotransmitter, which means that these cells are more likely to produce excitatory potentials in their postsynaptic targets. The flow of ionic current across the synaptic and extrasynaptic membranes is what produces excitatory postsynaptic potentials. An intracellular recording reflects the flow of excitatory postsynaptic potential current across both the membrane resistance and extracellular resistance, whereas an extracellular recording is measured across the extracellular resistance only. Since the extracellular resistance tends to be much smaller than the membrane resistance, the extracellular potentials are generally smaller than intracellular potentials. For this reason, as well as the fact that the size of potentials decreases as a function of distance from the source, it is nearly impossible to record the activity of a single neuron from the scalp (for a comprehensive review, see Nunez and Srinivasan 2005).

Therefore, electrical activity of large assemblages of neurons showing synchronous activity must be summated to produce electrical potentials that are sufficiently large to be recorded by EEG on the surface. While EEG recording reveals the workings of neuronal ensembles with exquisite temporal resolution, there are some key limitations. For example, given its extracellular nature, the potentials are always recorded on the surface, some distance away from the source generators. Several techniques have been established to narrow down the location of the source, including dipole source analysis; nevertheless, they are based on strong assumptions about how many sources may be generating any given signal and provide only an approximate location of these possible source generators. Further, although the recorded electrical potentials are modulated by synaptic activity, the potentials themselves do not provide information about events at the synaptic level. For instance, the polarity of the recorded scalp potentials depends on whether the postsynaptic potentials were excitatory or inhibitory, and whether the synapses were located in a superficial or a deeper layer. Therefore, the directional deflection of the extracellularly recorded potentials on the scalp does not inform us about the nature of the synaptic potentials. Nevertheless, they can provide meaningful information on neurocognitive processes, which require large-scale integration of neuronal activity.

Spontaneous Electroencephalographic and Event-Related Potential Data

Neural activity can be recorded during sleep or rest, and these recordings are sometimes referred to as spontaneous EEG. Applying an algorithm (namely the Fourier transform) to spontaneous EEG allows for the decomposition of the data in the frequency domain. This process relies on the basic assumption that the recorded EEG is relatively stable over the recording period. The range of frequencies typically observed from the human scalp usually spans 0 to 50 Hz. This can be further separated into various major frequency bands: delta (0.5–4 Hz), theta (4–8 Hz), alpha (8–12 Hz), beta (12–30 Hz), and gamma (>30 Hz). Activity in different frequency ranges has been differentially associated with wakefulness stages (Steriade et al. 1993) as well as neurocognitive functions (Klimesch 1999, Ward 2003).

Electrical activity in the brain can also be recorded during task performance. As we noted earlier, the part of the EEG time-locked to a stimulus or response is referred to as an ERP. To obtain a reliable estimate of neural activity, ERPs are generally averaged across many repeated presentations of an identical event to increase the signal-to-noise ratio. This way, randomly occurring (or non–event-related) activity or noise in the data is theoretically cancelled out (e.g., Perry 1966), revealing the characteristics of the averaged evoked potentials. Each ERP component, which indicates gross changes in electrical potentials, can be distinctive in shape, magnitude, and latency. The functional interpretation of an ERP component depends on all of these characteristics, as well as the site of recording and the event of interest.

Equipment

EEG data acquisition requires several pieces of hardware and software. Experimenters typically purchase turnkey systems that include the electrodes, cap, amplifier, battery box, and recording software, whereas the stimulus presentation software typically necessary for ERP-based studies is often purchased separately. This section first details the major pieces of hardware necessary for EEG data acquisition, namely the amplifier, electrodes, and cap. Subsequently, we introduce several important features of presentation software.

Electrodes

There are two major types of electrodes—"active" electrodes, which involve built-in circuitry that amplifies the electrical signal, and "passive" electrodes, which have no built-in circuitry. Active electrodes require less preparation and typically produce better signal quality than passive electrodes. Depending on which system is used, one can record data from a single electrode up to 32 electrodes and multiples of this number up to 256 electrodes. Most electrodes used for EEG studies are made of tin or silver covered with a silver chloride coating. The standard placement of EEG electrodes over the scalp follows the International 10/20 system, whereby the locations of electrodes are set at a standardized distance away from each other, and the nasion–inion (i.e., the point between the forehead and the nose, and the lowest point of the skull at the back of the head, respectively).

Amplifier

In starting an EEG study, one of the first and most important decisions to make is which amplifier to purchase. There are several characteristics that differentiate one

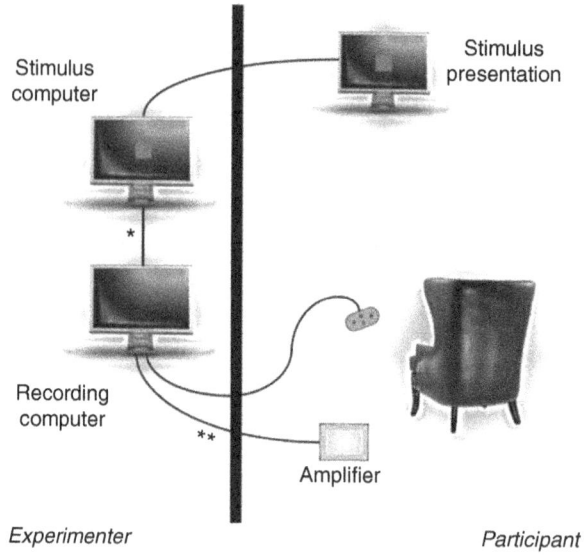

FIGURE 6.1 Setup of EEG equipment. The setup generally involves establishing a connection between the stimulus presentation software and data acquisition software (*), as well as a connection between the amplifier and the recording software (**).

brand of amplifier from another. First, some amplifiers are made for clinical use only, while others are for research purposes, which is what the current section will focus on. Second, most amplifiers today are digital instruments, which convert the analog EEG signal into a series of discrete numerical data points. The number of data points per second during the analog-to-digital conversion reflects the amplifier's sampling rate, which is expressed in Hertz (Hz) or samples per second. Importantly, the sampling rate determines the highest frequency of the analog EEG signal that can be faithfully represented in the digital time series data—only frequencies half the sampling rate or less. This cutoff, referred to as the Nyquist frequency, requires one to predetermine the highest frequency of interest in the given EEG application and set the sampling rate at two times or more above it. Lastly, for ERPs, the recording program records the signal from the amplifier and should also simultaneously receive and record codes that correspond to stimulus onset or responses made if participants were performing a task. Figure 6.1 presents the general setup of the major pieces of equipment.

Stimulus Presentation Software

The stimulus presentation software used in an EEG study is similar to that used in other cognitive-behavioral studies. The major difference is that as stimuli are presented, a separate trigger code corresponding to the stimulus onset or other events of interest must be sent simultaneously to the recording software so that the neural response can be time-locked to the event during signal averaging. Accordingly, one of the most

TABLE 6.1 Stimulus presentation software. This table compares and contrasts three commonly used stimulus presentation software packages.

	E-Prime	MATLAB	Presentation
Cost	Most expensive	Less expensive	Less expensive
Ease of programming	Mainly GUI-based	Mainly code-based	GUI and code
Flexibility	Moderate	Excellent	Excellent
Timing control	Very precise	Can be precise	Very precise
Stimuli	Easy to display visual and auditory stimuli, separately or simultaneously	Requires extra lines of code to present complex stimuli	Easy to display visual and auditory stimuli, separately or simultaneously
Support	Support system staff; user forums	Commercial guidebooks; user forums	Comprehensive documentation; user forums
Operating system	PC	PC or Mac	PC

important deciding factors in choosing a presentation system is the timing accuracy of stimulus presentation. Table 6.1 compares and contrasts three existing commonly used programs, E-prime, Matlab, and Presentation.

Nature of Stimuli. In addition to providing a timing marker, another issue is the nature of stimuli in the given task. If the stimuli are simple and unimodal, then most software can present them effectively. However, while most software packages are capable of presenting complex, multimodal stimuli (e.g., visual and auditory), the efficiency and convenience of presenting them vary across different software packages. This is especially an issue if the task requires synchronization of multiple stimuli across modalities; therefore, stimulus presentation is closely associated with programming of the task.

Support System. One useful feature that most software offers is online support. In addition to detailed manuals, there are several other modes of support available, including one-on-one email correspondence and online forum or chat groups. This may be a case of "you get what you pay for." In particular, the more expensive software packages often have support personnel who can provide a detailed response in a timely manner. However, this is not to say that less expensive software necessarily offers less support; in fact, due to its affordability, there are likely more users interacting online to provide support to each other. Especially for simple questions, you are likely to find a response among the pages and pages of forums and chat groups. The only caveat is that searching for an appropriate answer may be a time-consuming process, after which you will still have to validate the accuracy of the response yourself. One can also search for existing workable versions of the entire experimental task online, as some websites provide a free version of scripts programmed for commonly used experimental tasks.

Data Collection, Preprocessing, and Quantification

There are numerous options and parameters for EEG data acquisition, preprocessing, and quantification; choosing the appropriate options at each step requires thoughtful planning in advance. Generally speaking, for ERPs, we record the EEG while participants perform an experimental task. In this case, the onset of any stimuli and responses during the experiment are time-stamped during EEG recording, making it possible to align the EEG responses in time with their onset. The voltage changes within an epoch of EEG time-locked to the stimulus or response of interest make up the ERP. These epochs of EEG are then averaged across numerous stimulus presentations to derive an average ERP. Notably, EEG can also be recorded while participants are not performing any tasks but are resting or sleeping. The different processing strategies for ERP and resting EEG data are most apparent at the data quantification stage. The following section begins by describing the general protocol for data acquisition, followed by a description of the type of data obtained, as well as steps for data processing. Subsequently, we explain methods for quantifying both EEG and ERP data.

Data Acquisition

The setup of equipment prior to the recording generally involves preparing the electrodes, placing them on the participant, and connecting the electrode wires to the amplifier. The goal of these steps is simply to record whatever electrical activity is detected on the scalp of the participant. However, additional steps are taken to maximize the likelihood that the recorded activity is clean and free of noncerebral sources; for example, asking participants to remain still and relaxed during recording to minimize unwanted muscle-related artifacts, and to blink only during intertrial intervals, when important stimulus events are not occurring. We will focus our discussion mainly on the electrodes, as they require the most attention during equipment setup and for maintenance of recording quality.

Preparing Electrodes. Electrodes placed on the face around the eyes measure electro-oculograms (EOGs), whereas electrodes placed on the scalp measure EEGs. For most electrodes, soaking them in a saline solution for 5–10 minutes before recording facilitates the connection between the electrodes and the scalp. Prior to placing the electrodes on the participant, the facial areas on which electrodes will be placed for recording need to be cleaned, usually with a piece of alcohol wipe. Any substances on the face, such as makeup or sunscreen lotion, can potentially interfere with the connection. The experimenter will then typically insert electrolyte gel onto the electrodes, which facilitates the conduction of electrical signals from the scalp to the electrode. At this time, several individual electrodes are attached to facial areas around the eyes to measure EOGs, as shown in Figure 6.2. Depending on the choice of reference, additional electrodes may be placed on the nose, earlobes, or mastoid regions.

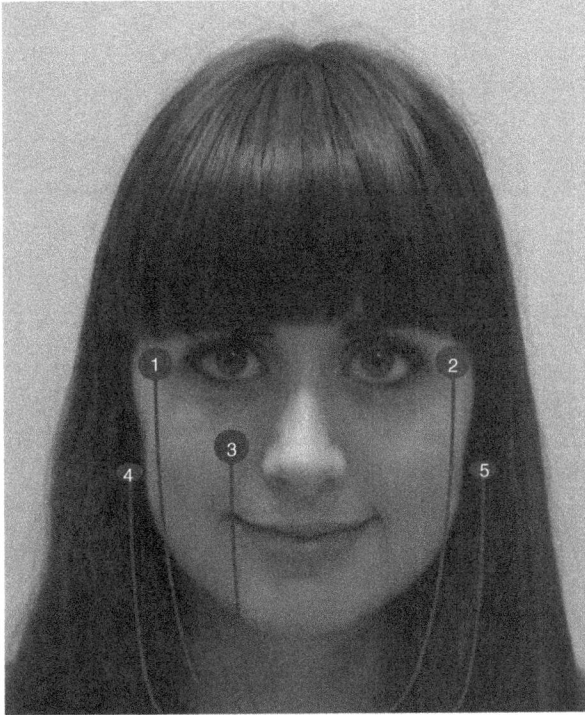

FIGURE 6.2 Placement of EOG electrodes. Horizontal EOGs are typically recorded from two electrodes placed on the right and left outer canthus of the eye (1 & 2). Vertical EOGs are recorded from two electrodes, with one placed inferior to the eye (3) and usually another one above the same eye measured by a frontal EEG electrode. Additional electrodes may be placed on both sides of the mastoids or earlobes (4 & 5), or on the nose, as reference electrodes.

Scalp EEG electrodes are typically attached to a cap to ensure that the electrodes are placed in a location that accurately corresponds to the labeled channels being recorded, and that this process can be replicated across participants. Therefore, it is important to ensure that the electrodes are attached to the correct location on the cap, and that the cap is appropriately placed on the participant's head with respect to the nasion and inion. Each scalp electrode is given an individual label such as F3 or P2, where the letter indicates the approximate lobe over which the electrode sits (F = frontal, P = parietal, O = occipital, and T = temporal) and the number indicates hemisphere (odd numbers = left hemisphere, even numbers = right); electrodes starting with a "C" are across the central sulcus, and electrodes ending with a "z" rather than a number run down the scalp midline. The standard placement for the electrode montage has electrode CZ (on top of the head) halfway between nasion and inion in the anterior/posterior direction and halfway between the two ear canals in the lateral direction, as shown in Figure 6.3. Subsequently, one must make sure that the electrical impedance with passive electrodes is within an acceptable range to ensure a high signal-to-noise ratio, with the range depending on the exact recording system used (i.e., some systems tolerate higher impedances

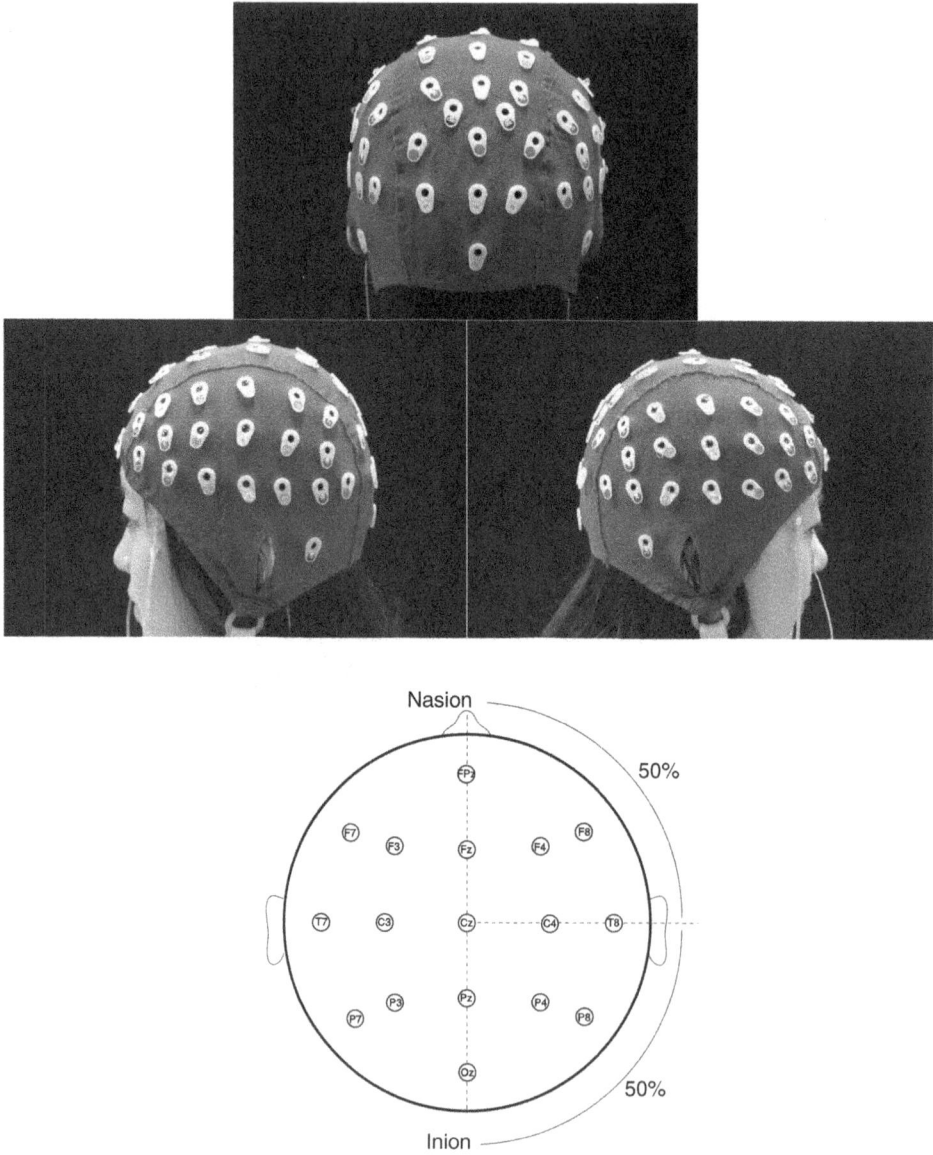

FIGURE 6.3 Placement of EEG cap. The standard placement has the vertex (i.e., Cz) halfway between nasion and inion in the anterior/posterior direction and halfway between the two ear canals in the lateral direction.

than others). On the other hand, EEG activity measured by active electrodes is much less susceptible to influences by impedance; therefore, the level of DC offset is used instead to index the quality of electrode contact. The amount of time required to perform these steps depends on the type of electrodes (e.g., active vs. passive), the number of electrodes (e.g., 32 vs. 128), and various factors that are often beyond the control of experimenters and intrinsic to subjects (e.g., sweat, hair thickness, and skull composition).

Equipment Maintenance. Most EEG electrodes are not inexpensive, so it is important to take good care of the electrodes and cap to maximize their lifespan. We describe several guidelines for maintaining each piece of equipment. Electrodes should be carefully treated during preparation and during cleaning after the experiment. The electrode ends of the cables should not come into contact with metal surfaces, and the other end of the cables should not come into contact with water. During the cleaning process, one can use a Q-tip and gently rinse the electrodes under warm water to remove gel residue. After all electrodes are thoroughly washed, they need to be dried completely and then stored in a dimly lit area. Similarly, the cap should be washed in warm water with soap until it is free of gel. To avoid stretching the cap, one can lay it flat to dry. The cleaning of electrodes and cap requires close attention not only for maintenance of the equipment but also for hygienic reasons, as they come into direct contact with the participants. It is highly recommended that researchers pay close attention to cleaning and storage details of their particular system.

Data Preprocessing

The raw data obtained from the amplifier contain the digitally converted set of numerical values that indicate the amplitude of the scalp potentials at each electrode site over time. Before statistical analysis, the data typically undergo several basic signal preprocessing steps, which are usually performed using the same software that is used for data quantification.

Rereferencing. During recording, EEG data are often referenced to one or two electrodes, which means that the selected reference channel is subtracted from all the EEG channels to eliminate nonneural effects and isolate neural effects. This step can be done online (i.e., during recording) or offline (i.e., after recording). The most typical reference channels used are located on the nose, earlobe, or mastoid regions. Another option is the average common reference, which involves subtracting the average of all EEG channels from each individual channel. Depending on the type of data to be analyzed, the choice of reference can change the resulting data considerably. While this is an ongoing debate, some have argued that the best available approach for high-density recordings (i.e., >60 channels) is the average common reference (Dien 1998). For further reading on choice of reference and the consequences of those references on EEG data, see Luck (2005), This is an especially important issue to consider when the purpose of the study involves examining the relationship between two electrodes (Nunez et al. 1997), which will be discussed in a later section.

Filtering. Digital filters (vs. analog filters) implement mathematical operations on the digital time-series data in order to increase or attenuate specified frequencies. There are several types and families of filters and numerous filter parameters. For example, a simple high-pass filter attenuates all data points below a certain cut-off frequency (e.g., 0.5 Hz), whereas a low-pass filter attenuates high-frequency data points (e.g., >60 Hz). High-pass filters can eliminate slow drifts and low-pass filters can eliminate distortions in the data. To preserve a midrange of frequencies, one can set a band-pass filter, only keeping data that fall within a specified frequency range. One can also run a notch filter to attenuate a narrow frequency range (e.g., 50 Hz or 60 Hz, to reduce AC line interference). The type of filter determines the mathematical function used to implement the process. Some of the commonly applied filters include Butterworth, Gaussian, infinite impulse response, and finite impulse response filters. A thorough discussion of each of these families of filters and their effects on the EEG signal is beyond the scope of this chapter; The interested reader is referred to Edgar, Stewart, and Miller (2005).

Artifact Detection and Correction. An equally important signal processing procedure is artifact detection and correction/removal. EEG data are highly susceptible to contamination by noncerebral signals. Eye blinks or saccades and muscle movements are the two most common types of such artifacts. Given their potential confounding effects on EEG data, it is important to detect and correct for them to accurately interpret the observed neural response. One can set varying criteria to determine the presence of an artifact. In the case of detecting ocular artifacts, these criteria are only applied to the EOG channels. For example, one can set an absolute threshold, whereby any segments of data exceeding a certain voltage in amplitude are flagged as an artifact. An alternative method is to set a peak-to-trough criterion, such that if the absolute difference between peak and trough within an epoch exceeds a predetermined value, it is considered an artifact. A more complex way to detect artifacts is using independent component analysis to decompose the sources of the EEG and EOG signals. From the decomposed sources or components, one needs to identify those that appear to be ocular in origin and then remove those sources from the data. Upon detection of artifacts in the traditional way (i.e., threshold-based detection), there are two options. Artifact correction subtracts the artifacts from the flagged segments, thereby altering the nature of the EEG data, while artifact removal completely deletes the selected data segments from subsequent analysis. These are only some of the most basic ways to process EEG data. The available data analysis software, whether commercial (e.g., Analyzer) or free (e.g., ERPlab), often determines the range of available options with respect to data preprocessing and analysis.

Data Quantification

After completing these preprocessing steps, the data are ready to be quantified. Different techniques are used to quantify and analyze resting EEG data and ERP data.

Resting EEG Data Analysis. Resting EEG data are recorded when participants are not performing a task, either with their eyes open or closed. The purpose of quantifying this type of data often includes generating measures of power in certain frequency bands and coherence between signals, and comparing these measures across different conditions and/or populations. Although no external task is assigned to participants during the recording of resting EEG, much evidence now suggests that "baseline" or "default" patterns of activity at various frequency bands correspond to different cognitive processes, like attention and memory. It is important to note that emerging techniques have evolved in recent years to examine power spectrum measurements in event-related EEG recordings as well—that is, measures of the power and coherence of activity time-locked to a stimulus or response.

First and foremost, as ocular artifacts can seriously contaminate measures of EEG, it is important to adequately correct for these artifacts in order to get a clean signal. Emerging evidence suggests that more complex techniques like independent component analysis are superior to traditional methods in eliminating artifacts (Keren et al. 2010, McMenamin et al. 2010, Shackman et al. 2010). With artifacts removed, one approach to analyzing resting EEG recordings is to quantify them by measuring the amplitude or power at each frequency averaged across time points. This can be done using the Fourier transform, which decomposes the complex EEG signal into a series of sine-wave components, characterized by the frequency, amplitude, and phase of the wave. However, fine temporal information about the original waveform is lost in this transformation. Figure 6.4 presents EEG power averaged over 2-sec epochs between 0 and 50 Hz at Cz.

In addition, resting EEG data can be characterized by coherence, which is the coupling between two different time series independent of their power. In essence, it captures the magnitude of the relationship between two distinct signals, as indexed by scalp electrodes, or independent components. The choice of reference and issues

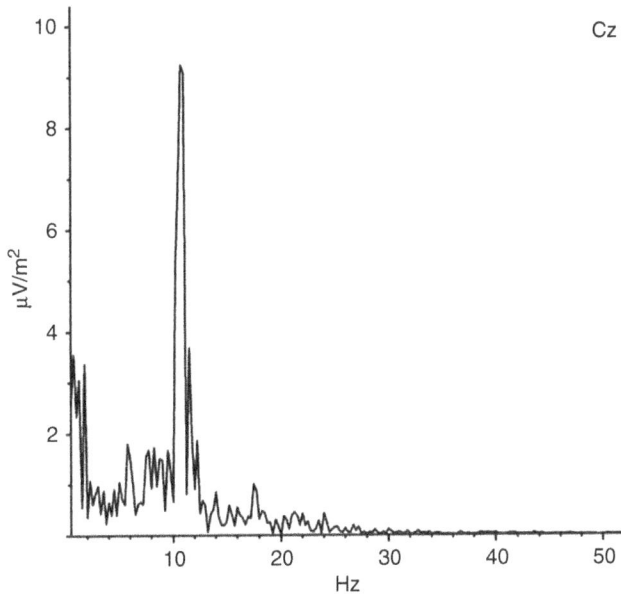

FIGURE 6.4 **EEG power.** EEG power is shown here across 0 to 50 Hz averaged across 2.048-s segments at Cz.

of volume conduction have been shown to confound coherence estimates on scalp electrodes. In particular, many of the common references used (e.g., nose reference, linked mastoid reference, or average reference) lead to erroneous coherence estimates (e.g., Nunez et al. 1997). Further, volume conduction effects (i.e., the transmission of an electrical signal from one source generator into various nearby electrodes) have also contributed to high coherence between neighboring scalp electrodes, which in reality may have received signals from the same source. One potential solution to these problems is to transform EEG data into current source density estimates, which is a reference-free measure that isolates the source of activity recorded at each electrode, thereby addressing both issues of reference and volume conduction effects (for detailed discussion of current source density and how it is calculated, see Nunez et al. 1997, Tenke and Kayser 2012).

ERP Data Analysis. ERPs are signal-averaged epochs of EEG activity time-locked to a stimulus or response. The functional goal of analyzing ERPs typically is to determine the extent to which sensory, cognitive, or motor responses to a stimulus vary between conditions and/or populations. These neural responses are represented by specific ERP components, which are positive- and negative-going deflections in the ERP waveform, such as the N100 and P300, where the letter indicates the polarity of the deflection (positive or negative) and the number indicates the approximate timing of the deflection onset/peak, after a stimulus (in milliseconds). The averaging of EEG epochs to derive ERP waveforms and components is portrayed in Figure 6.5. Analysis of ERP waveforms involves the quantification of data in the temporal, spatial, or temporospatial domains. Our discussion focuses on the temporal domain, which entails quantifying the data recorded at individual electrode sites and time-locked to a stimulus or response. Statistical comparisons can then be performed between experimental conditions. The two main considerations in quantifying an ERP component are choice of electrode sites and measured characteristics of the specific component(s) (e.g., latency and/or amplitude). For instance, one might observe a delayed P300 (i.e., an increase in latency) at parietal electrode sites in one condition, or a larger P100 (i.e., an increase in amplitude) at occipital sites in another condition. In the remainder of this section, we describe two common measures of an ERP component, amplitude and latency, and highlight some issues in data quantification.

One of the most common measures of an ERP component is its amplitude, and there are three approaches to this measurement. First, the peak amplitude is the amplitude at the time point when it has reached its minimum or maximum. The mean amplitude is the averaged amplitude across all time points within a specified time window. One can also compute the peak-to-peak amplitude, which measures the amplitude of one peak relative to an adjacent peak in the waveform. The choice of a peak versus mean amplitude measure depends on several factors, including the presence of noise, whether

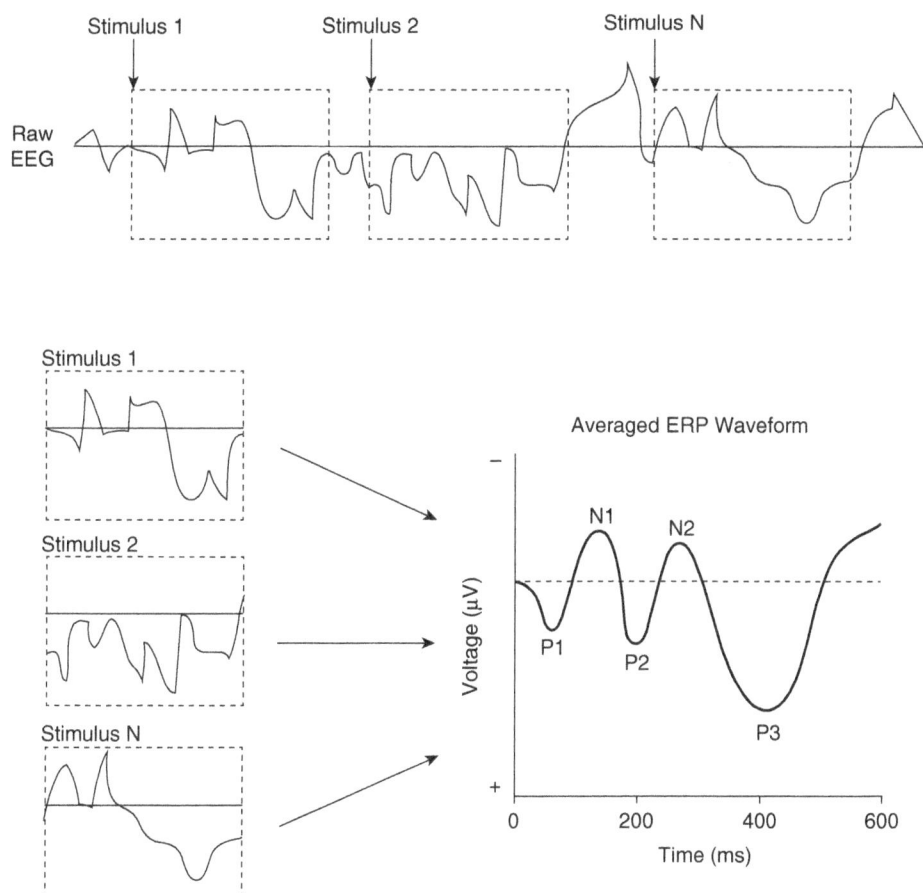

FIGURE 6.5 ERPs. Raw EEGs were segmented into equal-length epochs time-locked to the same stimulus (top, bottom left) and averaged to create an ERP waveform with several ERP components (bottom right).

an ERP component has a well-defined peak, and whether the latency of an ERP component varies between conditions.

The latency of an ERP component is another commonly used measure. This method of quantification is assumed to reflect the speed and relative timing of perceptual, cognitive, and/or motor processes. One approach is to identify the peak latency—that is, the time point at which the amplitude reached its maximum/minimum within a specified time range. Another approach is the onset latency, which is the time point after stimulus onset at which the component began (e.g., when amplitude deviates from baseline). However, in many instances it can be challenging to determine when exactly a component begins; one approach is to calculate line slopes or gradients near an obvious bend in the waveform and then select the time point of maximal change between slopes/gradients (e.g., Schwarzenau et al. 1998).

Regardless of the approach, this type of measure allows for comparison of temporal differences between experimental conditions. For a more detailed description, see Handy (2005).

In addition to careful planning and setting of hypotheses a priori, there are several issues to consider in the quantification of an ERP component. First, exclusion criteria must be set based on not only behavioral performance but also the characteristics of the neural response. Specifically, the absence of a component or the occurrence of a component outside of the expected time or at an unexpected electrode site may provide valid reasons for exclusion. Second, the comparison of waveforms across conditions may sometimes require one to set different parameters to accurately capture the property of interest in the two waveforms. For instance, if the measure of interest was mean amplitude, and the peak latency of an ERP component turned out to be earlier in one condition relative to a second condition, then it would be appropriate to choose different time windows that correspond to the peak latency of each waveform, and to consider latency as a relevant variable. A third point of consideration concerns the presence of noise in the data, which can be differentiated into random noise and systematic noise. Random noise reflects signals that are uncorrelated with the stimulus or response of interest—for example, heartbeats and interference from adjacent power sources. These signals will theoretically cancel out over the course of the recording. In contrast, systematic noise reflects signals that correlate with the stimulus of interest, such as eye blinks or microsaccades after every presentation, and thus would not average to zero in the waveform. Fortunately, there are well-established techniques to estimate and eliminate both random noise (Gratton et al. 1989) and systematic noise (Gratton 1998, Woldorff 1993).

EEG and ERP have been and will continue to be an important tool for understanding neurocognitive functions. For researchers interested in furthering their knowledge of these methods, the suggested readings below would be an excellent next step.

Frequently Asked Questions

Q: What are the major factors to consider when purchasing an EEG system?

A: The most important factor is likely to be cost. This includes the fixed cost to purchase a new system in the beginning, as well as the additional maintenance costs (e.g., to replace strands of electrodes). Another factor to consider is the type of electrodes (i.e., active electrodes or passive electrodes), and the number of electrodes (e.g., 32 vs. 128 electrodes), which will depend on the nature of your experiments. The purpose of the system and the population of interest are also points of consideration. For example, if the given research involves mostly infants, then the system of choice would be one that requires the least amount of time for electrode preparation on the participants.

Q: How do I decide on timing of stimuli during programming?

A: Previous ERP experiments will likely provide useful information to reference. Two important considerations are the expectancy effect sometimes observed in an ERP study and the possibility of overlapping responses. Specifically, if a stimulus is always presented at exactly the same interval, then an ERP component may occur just before the stimulus that appears to index one's expectation that a stimulus is about to occur. This may confound the actual stimulus-evoked response of interest to the experiment. Further, two stimuli of similar nature presented too quickly one after the other may result in overlapping ERPs that are difficult to dissociate.

Q: What are some biological and environmental factors that might affect the quality of recording?

A: Biological factors intrinsic to subjects include the thickness of their hair and skull, which is a medium of conductance. Other subject-related factors include drug intake and alcohol consumption prior to recording, and whether the subject was caffeine- or sleep-deprived during recording. Research has shown that individuals who have experienced a concussion or epilepsy exhibit EEG waveforms different from those who have not. Environmental factors include power lines, radiofrequency transmission from a nearby cellular phone or microwave, sudden loud noises, and so forth. To minimize environmental confounds, some experimenters prefer a sound-attenuating booth.

Q: How much data are necessary—specifically, how many subjects and how many trials in an average?

A: The number of subjects depends on the research design, research question, and size of the anticipated effect. As in any other experiment, within-subject designs are always more powerful than between-subjects designs and therefore would require fewer subjects. Importantly, a research question that concerns a difference in neural response between conditions would likely require fewer subjects (e.g., 25 per condition) than one that concerns the relationship between a personality trait and the magnitude of a neural response (e.g., 80–100). If data from previous research are available, then a power analysis can provide a sound estimate of the number of subjects required for each condition. Further, if the anticipated effects between two conditions are large, then fewer subjects may be needed to reveal the differences. There are no standard rules to follow for the minimum number of trials necessary to include in an average; rather, it is mostly a function of the signal-to-noise ratio, which once again may depend on the task and the quality of data. Nevertheless, some guidelines have been developed. For example, a minimum of 30 trials have been recommended for a well-defined P300 ERP component, whereas a minimum of 6 trials have been shown to be sufficient in generating the error-related negativity ERP component.

Q: Which of the basic preprocessing steps should be implemented?

A: Some of the basic steps include filtering and artifact detection. There are varying opinions on the need for filtering. In contrast, perhaps most would not argue that artifact correction/removal is a crucial step of data preprocessing.

Q: What are the guidelines for ERP data analysis, in terms of which electrodes to include, what time period to assess, and which measures to use?

A: That all depends. Generally speaking, all three of these parameters would ideally be set a priori based on previous experiments and the nature of the current study. First, the electrodes selected for analysis should reflect the location on the scalp where the component is expected. For example, in examining the sensory response to a visual stimulus, one would expect to observe the P100 at electrode sites over the occipital cortex, but the P300a ERP component should be maximal over the frontal/central midline sites. Second, the time period for analysis should be selected based on the characteristics of the observed component and latency at which the component is expected. Specifically, the time period assessed should capture the entire component of interest, but nothing more. Third, the choice of measure should be selected based on the nature of the research question. For instance, a research question that concerns the difference in the extent of cognitive processing between two clinical populations would primarily rely on an amplitude measure, whereas a study that examines the speed of processing would likely use a latency measure. However, both measures may provide unique and important information in each case (Table 6.2).

TABLE 6.2 Necessary Equipment

Equipment	Potential Sources
EEG hardware and software: Electrodes and cap, amplifier, battery box, recording software, and miscellaneous products (electrolyte gel, syringe, stickers)	Biosemi (www.biosemi.com)
Stimulus presentation software	E-prime (www.pstnet.com), MatLab (www.mathworks.com), Presentation (www.neurobs.com) (see Table 6.1 for a comparison between these three applications)
Parallel port device	Computer/electronic stores
Subject cleaning materials: towels, shampoo, hair dryer, etc.	Retail stores
Equipment cleaning materials: Q-tips, soap, disinfectant	Retail stores

Further Reading

Handy, T. C. (Ed.) (2005). *Event-related potentials: A methods handbook.* Cambridge, MA: MIT Press.

Luck, S. J. (2014). *An introduction to the event-related potential technique; second edition.* Cambridge, MA: MIT Press.

Picton, T. W., Bentin, S., Berg, P., et al. (2000). Guidelines for using human event-related potentials to study cognition: recording standards and publication criteria. *Psychophysiology* 37(2): 127–152.

References

Dien, J. (1998). Issues in the application of the average reference: Review, critiques, and recommendations. *Behav Res Meth* 30: 34–43.

Edgar, J. C., Stewart, J. L., & Miller, G. A. (2005). Digital filters in ERP research. In T. C. Handy (Ed.), *Event-related potentials: A methods handbook* (pp. 85–114). Cambridge, MA: MIT Press.

Gratton, G. (1998). Dealing with artifacts: The EOG contamination of the event-related potential. *Behav Res Meth* 30: 44–53.

Gratton, G., Kramer, A. F., Coles, M. G. H., & Donchin, E. (1989). Simulation studies of latency measures of components of the event-related brain potential. *Psychophysiology* 26: 233–248.

Handy, T. C. (2005). Basic principles of ERP quantification. In T. C. Handy (Ed.), *Event-related potentials: A methods handbook* (pp. 33–55). Cambridge, MA: MIT Press.

Heinze, H. J., Mangun, G. R., Burchert, W., et al. (1994). Combined spatial and temporal imaging of brain activity during visual selective attention in humans. *Nature* 372: 543–546.

Keren, A. S., Yuval-Greenberg, S., & Deouell, L. Y. (2010). Saccadic spike potentials in gamma-band EEG: Characterization, detection and suppression. *Neuroimage* 49: 2248–2263.

Klimesch, W. (1999). EEG alpha and theta oscillations reflect cognitive and memory performance: A review and analysis. *Brain Res Rev* 29: 169–195.

Martin, J. H. (1991). The collective electrical behavior of cortical neurons: The electroencephalogram and the mechanisms of epilepsy. In E. R. Kandel, J. H. Schwartz, & T. M. Jessell (Eds.), *Principles of neural science* (3rd ed.) (pp. 777–791). New York: McGraw-Hill Professional Publishing.

McMenamin, B. W., Shackman, A. J., Maxwell, J. S., et al. (2010). Validation of ICA-based myogenic artifact correction for scalp and source-localized EEG. *Neuroimage* 49: 2416–2432.

Nunez, P. L., & Srinivasan, R. (2005). *Electrical fields of the brain: The neurophysics of EEG* (2nd ed.). New York: Oxford University Press.

Nunez, P. L., Srinivasan, R., Westdorp, A. F., et al. (1997). EEG coherency I: statistics, reference electrode, volume conduction, Laplacians, cortical imaging, and interpretation at multiple scales. *EEG Clin Neurophysiol* 103: 499–515.

Perry, N. W. (1966). Signal versus noise in the evoked potential. *Science* 153: 1022.

Schwarzenau, P., Falkenstein, M., Hoorman, J., & Hohnsbein, J. (1998). A new method for the estimation of the onset of the lateralized readiness potential. *Behav Res Meth* 30: 110–117.

Shackman, A. J., McMenamin, B. W., Maxwell, J. S., et al. (2010). Identifying robust and sensitive frequency bands for interrogating neural oscillations. *Neuroimage* 51: 1319–1333.

Steriade, M., Nunez, A., & Amzica, F. (1993). Intracellular analysis of relations between the slow (less-than-1 Hz) neocortical oscillation and other sleep rhythms of the electroencephalogram. *J Neurosci* 13: 3266–3283.

Tenke, C. E., & Kayser, J. (2012). Generator localization by current source density (CSD): Implications of volume conduction and field closure at intracranial and scalp resolutions. *Clin Neurophysiol* 123(12): 2328–2345.

Woldorff, M. G. (1993). Distortion of ERP averages due to overlap from adjacent ERPs: Analysis and correction. *Psychophysiology* 30: 98–119.

Ward, L. M. (2003). Synchronous neural oscillations and cognitive processes. *Trends Cogn Sci* 7: 553–559.

Electroencephalography and Local Field Potentials in Animals

Sean Reed, Sonia Jego, and Antoine Adamantidis

Introduction

This chapter discusses the history, practice, and application of electroencephalography (EEG) and local field potential (LFP) recordings, with a particular focus on animal models. EEG and LFPs are two of neuroscience's oldest, easiest to implement, and most robust tools for characterizing brain activity. Due to their similar biophysical basis, the terms *EEG* and *LFP* are sometimes used interchangeably, despite their differences in application and interpretive limitations. In practice, EEG is recorded by placing electrodes on the scalp or surface of the skull, providing a noninvasive survey of global brain dynamics. In contrast, LFPs are recorded by placing an electrode within neural tissue, providing a measurement of local electrical activity within specific brain regions. Both techniques involve recording the aggregate electrical activity from several thousands of neurons at once to determine how these neurons synchronize during various behavioral states. These classical techniques remain useful in both clinical and basic research settings, especially in the study of sleep and epilepsy, due to their applicability in both humans and other mammalian model organisms. The biggest advantage of EEG and LFP techniques is their ability to directly sample synchronous neuronal activity with near-perfect temporal resolution. While the fundamental principles behind these techniques have not changed, their instrumentation, analysis, and application have become much more sophisticated over time.

Since the development of the first EEG recording technique by Berger (1929), EEG/LFP technology has revolutionized daily clinical neurology and neurosurgery, from identifying aberrant LFPs in epilepsy patients to characterizing the normal changes in EEG activity associated with various behavioral states in healthy subjects. In recent years, refined imaging technologies with higher spatial resolution, such as functional magnetic resonance imaging (fMRI) or positron electron tomography (PET), have come to supplement or replace direct neurophysiological recordings of population activity, but EEG/LFP techniques continue to provide affordable real-time acquisition of direct neuronal activity in cases when it is necessary or useful. Importantly, the temporal resolution of EEG/LFP (milliseconds) is much greater than that of fMRI or PET (5–10 s), making these techniques the method of choice when a high degree of temporal resolution is needed.

Brain Rhythms

The EEG/LFP signal is a reflection of global state-dependent activity in the brain and is therefore sensitive to various behavioral and arousal states. The recorded traces can be separated into a set of sinusoidal frequency components using mathematical algorithms like the Fourier transform. The predominant constitutive components are what are commonly referred to as brain rhythms or brain waves, and are characteristic of specific cognitive or behavioral states.

Brain rhythms are commonly classified based on the frequency and amplitude of the dominant sine-wave component. Frequency is measured in cycles per second, while amplitude denotes the vertical displacement of the wave from the zero mark during one frequency cycle (Fig. 7.1a). Amplitudes are typically in the range of 0.5 to 150 µV in human EEG, but the EEG amplitude is much higher in rodents due to the proximity of skull screws to the brain tissue (~0.6 mV). Signals derived from LFPs are much larger in amplitude compared to EEG (>0.5 mV in rodents) because the electrodes are situated within active neuronal tissue and the currents do not have to travel through bone and muscle. Indeed, one of the major differences between EEG and LFPs lies in the subset of signals that they sample. Although they are generated from the same overall pool of electrical activity, the distance of the electrode locations (scalp vs. intracranial), from the field source will generate different types of signals. As a result, EEG signals do not provide reliable information regarding brain rhythms higher than ~40 Hz, whereas LFP can acquire up to ~200 Hz.

Brain rhythms are typically classified according to the following criteria:

Delta rhythms (δ, 0.5–4 Hz) are the predominant waveforms during slow-wave sleep (also called non-rapid eye movement [NREM] sleep). Delta waves during slow-wave sleep have a relatively large amplitude (75–200 µV in humans and up to 800 µV in rodents).

Theta rhythms (θ, 4–9 Hz) are more obvious in rodents than in humans and exhibit a characteristic "sawtooth" waveform on the EEG. Theta waves occur during various types of activities during wakefulness but are most consistently present during rapid eye movement (REM) sleep or paradoxical sleep (Jouvet 1969).

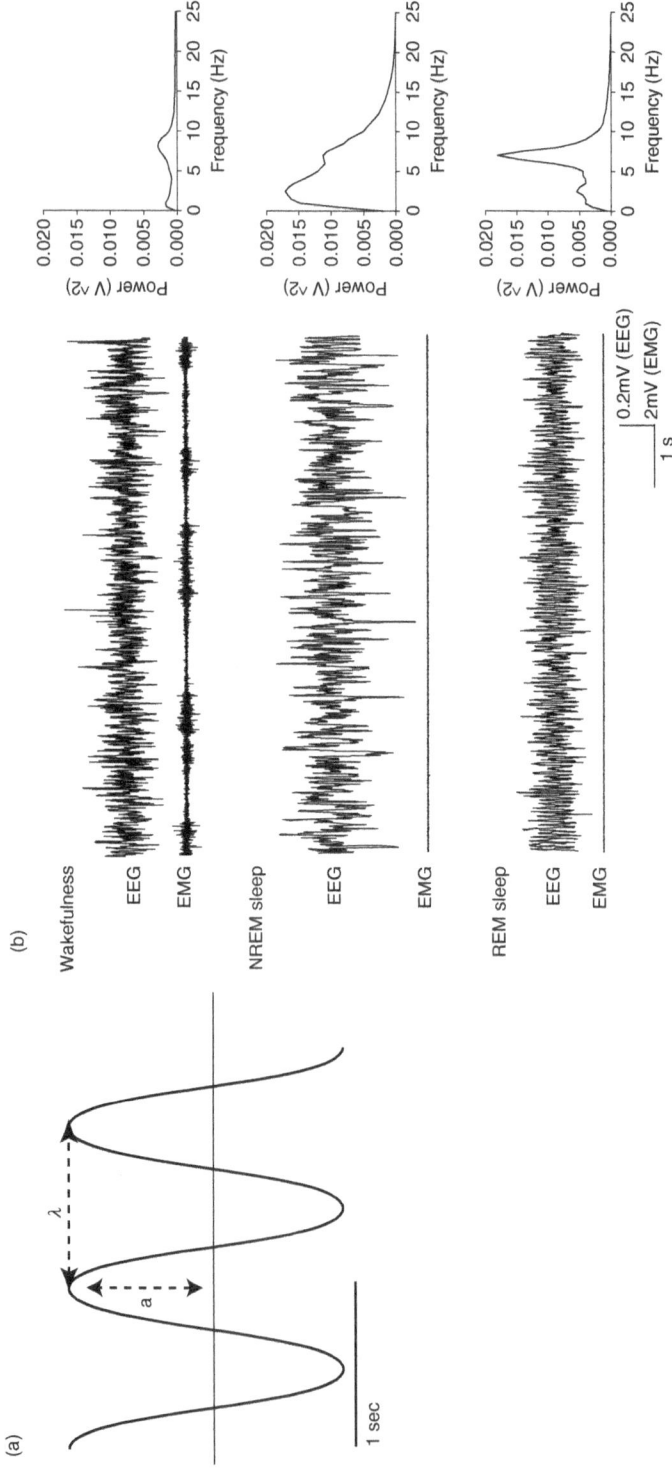

FIGURE 7.1 Characterization of brain rhythms across sleep/wake states. a: Brain waves are characterized by their frequency (in Hertz) and amplitude (a; in millivolts). b, left: EEG, LFP, and electromyographic (EMG) waveforms across the sleep–wake cycle of the mouse. Wakefulness is defined by low-amplitude, high-frequency (i.e., short-wavelength) EEG/LFP with high muscle tone. During NREM sleep, there is a slowing of the EEG/LFP, with a predominance of low-frequency components and increased peak amplitude. Constituent wavelengths in an EEG/LFP sample can be evaluated using power spectrum analysis, which indicates their relative contribution to the complex waveform (b, right). During NREM there is a strong delta (0.5—4 Hz) component and a marked reduction of the EMG. REM sleep is typically characterized by theta activity (6—9 Hz) as illustrated in the power spectrum, along with rapid eye movements and muscle atonia.

Alpha rhythms (α, 9–15 Hz) are predominant during wakefulness and are stronger when the eyes are closed and during relaxation. Alpha waves are more extensively studied in humans than in rodents and are preferentially observed in the posterior and occipital regions of the brain, with typical amplitudes of about 50 μV.

Beta rhythms (β, 15–30 Hz) occur during states of increased alertness and attention. These are perhaps the least studied and represented brain rhythms in the literature.

Gamma rhythms (γ, 30–200+ Hz) are fast waves present during focused arousal and are likely to reflect a global state of alertness. Because of the relatively high frequency of these waves, they are not observable using EEG and are typically studied using LFP.

In EEG, recordings are obtained from multiple topographical locations across the scalp simultaneously, providing a measure of the extent to which brain rhythms occur in synchrony across multiple regions (Fig. 7.1b). Some rhythms show strong coherence over the entire scalp (e.g., the delta activity), while others are topographically restricted (e.g., theta, gamma, beta). Therefore, the EEG signals recorded between two electrodes are highly dependent on electrode placement and are always dominated by lower frequencies relative to higher ones.

LFP signals are drawn from an extremely small brain region, so they cannot pool information beyond a relatively small diameter region around the electrode. As a result, LFP waveforms cannot provide any information concerning the topographical dynamics of synchronized activity across the brain. Instead, they provide a very focused profile of synchronized activity in microcircuits of the brain region where the electrode is implanted.

Fundamental Biophysics of Electroencephalography/Local Field Potentials

The underlying mechanisms of electrophysiological recordings are based on the collective ionic exchanges that occur between the intracellular cytosol and extracellular fluid within the brain tissue. The collective summation of all ionic currents from excitable membranes generates electrical fields that propagate through various media to reach the recording electrode. For EEG acquisition, electrical signals over the entire brain must traverse the extracellular space, cerebrospinal fluid, skull, and scalp to eventually converge upon the electrode. As a result of the inherent low-pass-filtering properties of extracellular media, noninvasive EEG provides a broad survey of superficial cortical activity with a 1- to 10-cm spatial resolution and a frequency resolution up to 40 Hz.

In contrast, LFPs are subject to far less filtering and therefore provide a much richer, although spatially restricted, signal. The spatial resolution of LFPs remains controversial and varies depending on recording location and parameters, with estimates ranging from a 250-μm radius (Katzner et al. 2009) to a radius of several millimeters (Nauhaus et al. 2009).

Postsynaptic potentials in neurons are thought to play the most significant role in generating the EEG/LFP signal, although glial cells have been suggested to partially contribute as well (Bedard et al. 2006, Birbaumer et al. 1990). To appreciate the biophysical basis of EEG/LFP recordings, it is useful to briefly review the neural basis of membrane potentials in neurons. Membrane potentials (V_m) are established by virtue of differential ionic concentrations between the extracellular fluid and intracellular cytosol separated by insulating selectively permeable bilipid cellular membrane. Whereas there is an abundance of potassium ions within the cell, there is an excess of sodium, calcium, and chloride ions in the extracellular fluid (Fig. 7.2a). As a result, an electrochemical gradient is established across the cell membrane that can be quantified as the membrane potential of the cell: $V_m = V_{in} - V_{out}$. Subthreshold currents are generated through the operation of permeable ion channels that exploit these electrochemical gradients and yield the influx of ions, thus depolarizing V_m (sodium influx) or further hyperpolarizing the neuron (e.g., chloride influx). Excitatory and inhibitory postsynaptic potentials (EPSPs and IPSPs) are the fundamental basis of acquired EEG and LFP waveforms (see Nunez and Srinivasan 2005 for in-depth biophysical details). Following ion channel activation, the subsequent inward or outward currents elicit local changes in the membrane potentials, which passively conduct toward the soma, thereby eliciting an action potential if the resulting V_m change surpasses a threshold (see Fig. 7.2b). Because of their passive nature, the recorded voltage disturbances from postsynaptic currents undergo exponential decay across distance.

An influx of current (sink) due to an EPSP in the dendrite is met with a nearly simultaneous efflux of passive current along the cell (source). This generates a dipole field along the dendrosomatic axis of the typical cortical pyramidal neuron. These rapid field changes across the membrane, as well as changes along the dendrosomatic axis, are detected by the electrode as a change of potential. It is worth noting that compensatory sink/sources do not occur instantaneously, and resulting monopoles may contribute to recorded potentials (Destexhe and Bedard 2012, Riera et al. 2012). Importantly, the magnitude of a field's contribution to EEG/LFPs depends on a number of factors, including the number and type of postsynaptic potentials generated, the degree to which activity is synchronous in a given location, and neuronal geometry.

Indeed, the geometric configuration of neurons in the vicinity of the recording electrode plays a critical role in the generation of field potentials and the ability to detect them. Cortical pyramidal neurons have a vertical arrangement of neuronal processes (i.e., dendro-axonal axis) that allows for a spatial separation of sinks and sources. The resulting dipoles have an *open-field* configuration that promotes current

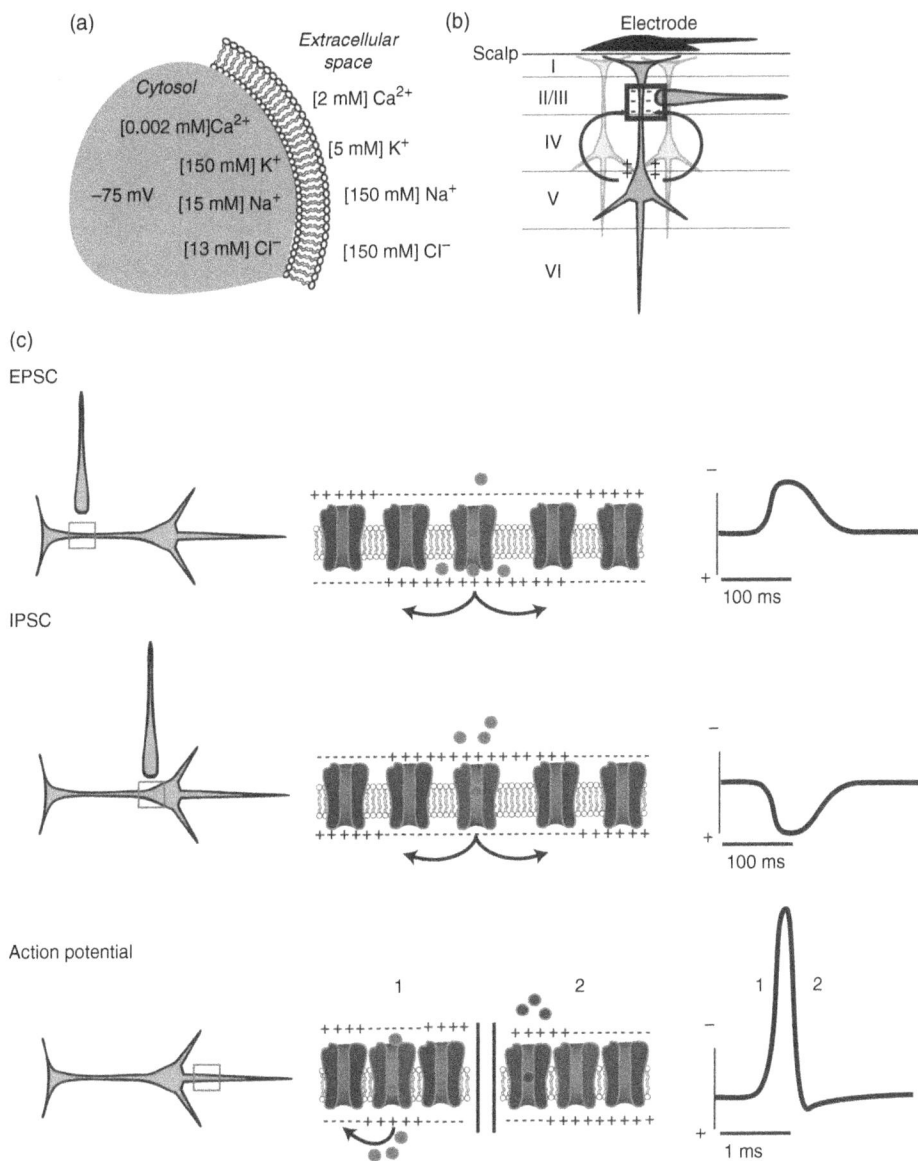

FIGURE 7.2 Mechanisms of field potential generation within the brain. a: Ionic concentrations within cytosol and extracellular space, separated by an impermeable phospholipid bilayer. b, top: Excitatory input in the form of EPSPs causes a localized influx of positive ionic current (sink), resulting in a negatively charged area in the extracellular space. An expanded view of this can be seen in c, where this is matched with a near-instantaneous efflux of current (source) in the soma resulting in a positively charged extracellular component. b, middle: The inverse (IPSPs). Together, IPSPs and EPSPs generate a dipole moment around the neuron, illustrated by the black arrows in c. Electrodes positioned perpendicular to the dipole detect the field closest to its position. Thus, an EPSP from superficial layers would generate a negative deflection on an EEG recording, denoting excitatory input. b, bottom: Action potentials do not significantly contribute to EEG field detection due to its extremely short time course (1 ms) compared to excitatory/inhibitory post-synaptic currents (EPSCs/IPSCs) (~100 ms). This shortened time course is attributed to the stepwise activation of sinks/source over short distances. Note that the activation of a sink (1) causes quick depolarization of the membrane, leading to the activation of a source (2), whereby the cell becomes repolarized in <1 ms. This process occurs multiple times as an action potential travels to synaptic terminals. In contrast, EPSCs/IPSCs lead to a passive diffusion of current, resulting in a long time course.

spread between these sink/source points across an individual neuron. In addition, the strict laminar cytoarchitecture and uniform dendritic orientations of pyramidal neurons allow for a spatial summation of many fields to generate an observable EEG trace, whereas the LFP electrode will simply survey the spatial summation across a more restricted space (see Fig. 7.2c). Conversely, *closed-field* configurations, which can be observed in thalamic neurons, do not have strict laminar structures and have many radial dendrites that extend in opposite directions. In this case, fields from spatially opposing dendrites are canceled, thereby yielding a weak observable extracellular field. Therefore, uniform cytoarchitecture (e.g., cortical pyramidal neurons) provides for open-field configurations that summate and yield the largest observable signal from dendritic postsynaptic potentials. The advantage of LFPs is that small potentials can be detected in brain structures that are devoid of strict laminar orientations, such as the thalamus. Although the field signal strength will be attenuated compared to that of the open-field cortex, closed-field orientations are not as important to LFP signal acquisition as they are in EEG.

However, open-field orientations and laminar alignments are not sufficient conditions for postsynaptic potentials to generate an observable EEG. For example, the cerebellum generates very weak EEG amplitudes, despite the fact that Purkinje cells are organized in a strict laminar fashion and have a vertical geometry similar to that of cortical pyramidal cells. The observed disparity between cortical and cerebellar recordings is thought to result from insufficient phase-locked synchrony and complex dendritic physiology in the cerebellum (Dalal et al. 2013). Indeed, temporal synchrony is essential for dipole summation as it elevates the generated extracellular field above the background noise. Pyramidal neurons in the cerebellum have >3,000 synapses (Rocher et al. 2008) and are constantly bombarded with innumerable postsynaptic potentials. In contrast, in the cortex, large-network synchrony via thalamocortical reciprocal loops, can orchestrate simultaneous EPSPs across cortical cells, thereby amplifying extracellular fields to detectable levels with EEG electrodes (for review see Siegel et al. 2012).

Despite these open configurations, not all electrical conductances can be readily detected on EEG or LFPs. Indeed, action potentials do not satisfy the biophysical parameters that produce detectable waveforms. Despite the extremely large fluctuations in V_m and current, these are stepwise events and cause only small portions of the entire axon to be depolarized at any one time. Moreover, due to the intrinsic low-pass-filtering properties of dendrites (Linden et al. 2010) and heterogeneous extracellular media (Bedard et al. 2004), detectable spikes from action potentials steeply decline with distance from the soma (Buzsaki and Kandel 1998, Henze et al. 2000). Secondly, an action potential is extremely short-lived (~1 ms), equivalent to a dominant frequency of 1 kHz. Postsynaptic potentials are much longer, lasting for up to 100 ms. Nevertheless, the relative proximity of LFP electrodes to active neurons has led physiologists to investigate the question of whether the characteristics of LFP waveforms may offer clues into firing behavior.

Two waveform characteristics found in both EEGs and LFPs are positive or negative deflections, depending on whether the waveform is oriented upward (positive) or downward (negative). It is assumed that these orientations reflect whether cells are depolarized), or hyperpolarized, but the directions depend on recording conventions and electrode orientation with respect to the generators. Figure 7.2c illustrates a population of cortical pyramidal neurons, which generally receive excitatory input on their superficial apical dendrites in layers 2/3 and perisomatic inhibition. Upon EPSP generation, the sink generates an extracellular negative field where the positively charged ions have entered the neuron. Conversely, the source generates a positively charged environment in the extracellular space where the current then emerges from the neuron. As a result, the outside component of the generated dipole is opposite to that within the cell. In contrast, an IPSP along the soma would result in a similar vertical dipole of opposite polarity. Thus, synchronous generation of postsynaptic potentials across a population of neurons would summate and amplify the field detection. Due to their strict vertical (i.e., laminar) anatomical arrangement and their dendrosomatic axis, dipoles in lissencephalic animals are all assumed to be perpendicular to the horizontal position of the EEG electrode. Electrodes will detect the polarity of the field that is closest to the cortical surface. Therefore, superficial EPSPs will generate a negative deflection and IPSPs will promote positive deflection. For this reason, it is convention to invert the waveform during EEG acquisition to avoid confusion.

Recording Techniques

The acquisition of EEG/LFPs requires the use of a transducer to convert bioelectricity into currents (voltage changes) that can be recorded and filtered using acquisition hardware (Fig. 7.3). In humans, scalp electrodes are used as the interface of this transduction (see Chapter 6), and voltage changes must travel through ~1–1.2 cm of skull and scalp. In animals, simple jeweler's screws are typically used as EEG electrodes and are implanted into the skull, thereby mitigating some of the filtering that occurs when electrical fields travel through insulating bone. A small hole (~300 microns in diameter) is drilled in the skull to position the metallic screws that are in contact with the cortical surface (see Fig. 7.3). Due to the screw's properties and its direct contact with the brain, amplifying conductive gel is not necessary. Each screw is directly connected to an amplifier via a conductive metal wire soldered to the screw. Thus, the animal is tethered during the recording procedure. Note that wireless telemetry devices have been successfully used in this configuration, although conventional wire-based connections still remain the most popular and cost-effective option.

As stated above, the quality of EEG data is contingent upon the source of the field that is close to the electrode. The experimenter must therefore ensure the correct placement of the electrodes in the area of interest. In human research, an international "10–20 system" of standardized electrode placements ensures reproducibility across

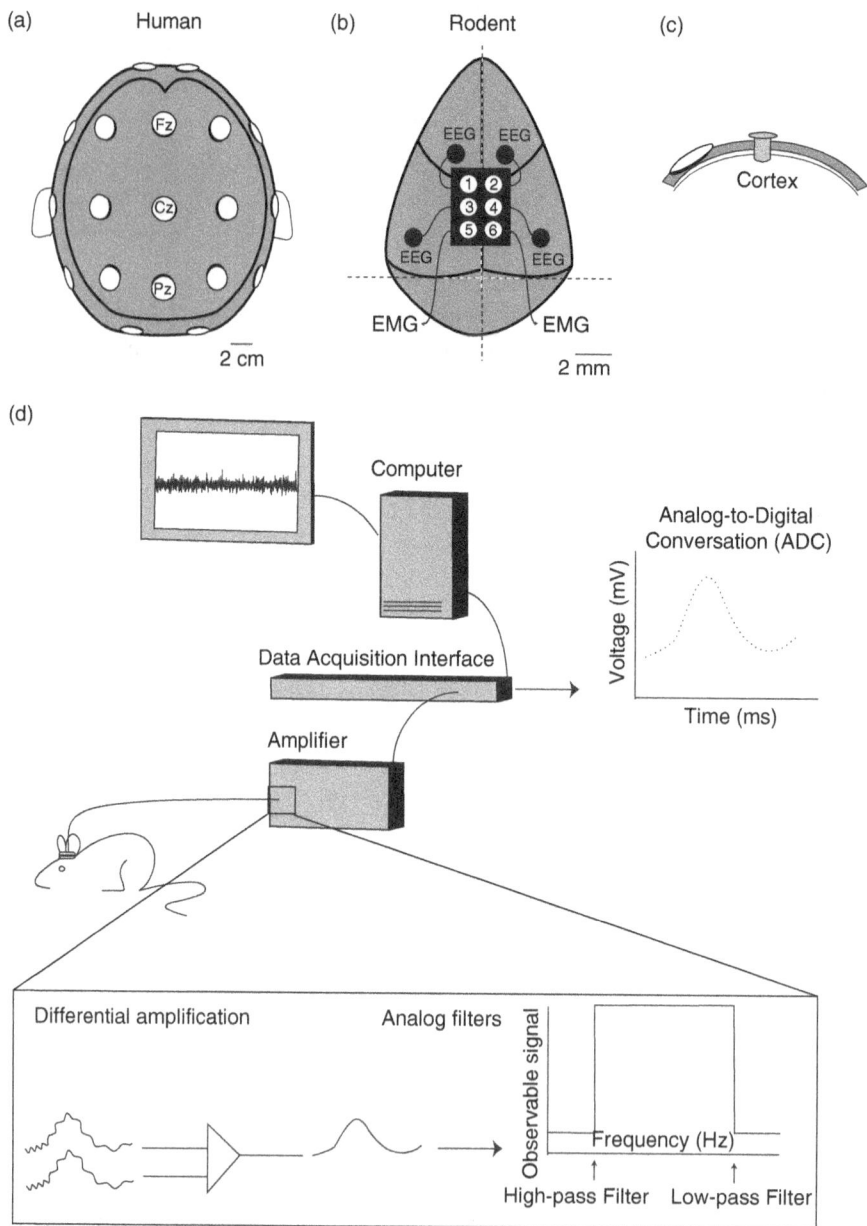

FIGURE 7.3 Comparison of human and animal recording techniques. a, left: In traditional human EEG recording, electrodes are evenly spaced and placed relative to identifiable locations around the skull (Fz, Cz, and Pz for example, illustrate center positions). In rodents, EEG screws are placed within the skull and may make contact with neural tissue and a head assembly is required for all the electrode wires to converge onto a plug that is attached to the amplifier. a, right: Side-view comparison of EEG electrodes in both human and animals, and LFP electrodes in animals. Note the extremely small contact size of LFP electrodes compared to the EEG screws as shown in the magnified view of the electrode tip (red box). b: With both EEG and LFPs, the first step in acquisition involves differential amplification that effectively removes any signal common to both inputs. Here we see the removal of 60-Hz power line noise. This is followed by a series of high- and low-pass filters that remove components of the signal not needed for analysis. Increasing the gain on a signal is an additional option (not shown), which effectively amplifies the signal for better detection. The signal is then sent to a digital acquisition interface, where it is digitized for storage and further analysis. This interface may also be used for generating stimuli and a host of other more complex arrangements that are beyond the scope of this chapter. Finally, the signal is stored on a computer for viewing and further online or offline analysis.

laboratories and publications (Jasper 1958; Chapter 6). The small number of electrodes present on the rodent skull establishes only very general estimates of field generation.

Techniques

Numerous different forms of electrodes can be used for LFP recording. Despite the large number of electrode geometries and impedances available, it has been argued that this does not significantly alter the recorded waveform (Nelson and Pouget 2010).

Single bipolar electrodes can be manufactured from inexpensive, commercially available nylon- or Teflon-coated stainless-steel wire (100 μm in diameter) with ~125 μm of the tip exposed and an intertip distance of ~0.5 mm. Two lengths of wire can be glued together longitudinally using epoxy adhesive, the insulation removed from the ends, and a gold connector soldered to one end of each wire. To fabricate multiple wound-wire electrodes at once, cut ~60 cm of wire and tie the ends together to form a loop. Attach one end of the loop to a hook and the other to a rotary device that will allow for quick winding of the wire (e.g., Dremel). Make the wire strands taut, mark off sections twice the desired length of electrodes, and place very small portions of tape along one strand of the loop at intervals twice the desired electrode length (e.g., 2 inches). The tape will prevent tight winding at the points where it is applied and will indicate where to cut the wound wire into separate pairs of LFP electrodes. Knowing when to stop winding takes some experience, since making it too tight will not allow one to adequately adjust the tip separation, and leaving it too loose will allow the strands to unwind. Cut the wires at the center of the tape. Carefully removing the tape will reveal loose ends. Use a flame from a lighter or other source, heat a very small portion of each cut end until the steel becomes red, and carefully remove the blackened Teflon from that portion with a razor blade. Gold-plated Amphenol pins can now be crimped to each of the four ends, and electrode assembly can be cut at the center to yield two bipolar electrodes for every segment between the tape. The last step is to separate the tips. Although there is no standard for exposed tip length or intertip separation, it is advisable to keep the separation <1.0 mm.

Artifacts

Despite the increasing complexity and specificity of EEG/LFP electrode arrangements, there is a realistic risk of trace contamination. Recordings are prone to contamination by electrical noise originating outside the brain, as well as technical contamination from power cables (60 Hz US/50 Hz Europe), poor cable connections, movement of the electrode, and so forth. The most common artifacts encountered in animal LFP/EEG recordings are high-frequency myogenic contamination, mastication, cardiorespiratory signals, and power line noise. Beyond the nuisance these signals present to researchers and diagnosticians, artifacts may act subtly to create the illusion of normal EEG (Yuval-Greenberg et al. 2008) or even epileptiform activity (Tatum 2013). Unfortunately there are no quick solutions to mitigate this contamination. Identifying artifacts and finding a suitable ground to increase the signal-to-noise ratio comes with experience (but see Chapter 6). Each laboratory setup is different and subject to different sources of interference. Advances in the design of acquisition hardware have aided in the automatic removal of extracerebral signals. Some artifacts, particularly myogenic ones, can be significantly reduced in LFP recordings through the use of bipolar differentials. Having two electrode contacts in close proximity (~100–900 μm) means that significant signal contamination in one contact will very likely exist in the other. Using sophisticated signal-processing techniques, one can selectively remove any signals common to both contacts, thereby mitigating any major artifact presence. After subtracting artifact signals, a bipolar electrode assembly will yield two different waveform channels devoid of signals that are common between the two contacts.

Amplification

Amplifiers increase the gain of the signal and can be used to measure the potential difference between two electrodes to remove any common signals between them (called the *differential*). This ensures that any electrical contamination present between two electrodes is differentially subtracted, thereby presenting a unique EEG/LFP trace (see Fig. 7.3). In animal EEG recordings, amplification typically occurs at a remote location 2–3 m from the screws on the animal's head. In animal LFP recordings, a ground electrode is placed on the animal as far from the active electrodes as possible, and the wires converge on a headstage mounted on the animal's head (see Fig. 7.3). For cortical recordings, the ground electrode is often placed in the cerebellum. The headstage to which the electrode is connected often contains preamplifiers, where initial amplification of the signals takes place.

Following amplification, the signal proceeds through a series of low/high-pass analog filters that restrict EEG/LFP frequency components within a defined window. For example, most sleep research focuses on changes that occur within 0.1–40 Hz. Therefore, one would use a low-pass filter to attenuate frequency components >40 Hz and a high-pass filter set to 0.1 Hz. LFPs record 1–200 Hz, so it may be useful to add a notch (or bandstop) filter, which admits or removes a specific band of frequencies from a trace. This provides an additional measure against unwanted 50- or 60-Hz power line noise. Commercially available LFP electrodes are often manufactured so that one can also derive action potentials from small populations of neurons from the LFP waveform by simply increasing the low-pass filter cut-off on the amplifier (e.g., 300–3,000 Hz).

After amplification and filtering, the waveform must be converted to a digital signal through an analog-to-digital conversion (ADC) board. Digitization (or sampling) is a fundamental component of EEG/LFP acquisition that will determine the overall fidelity and usefulness of the acquired trace (see Chapter 6). Digitization involves converting an analog time series into discrete time points at a constant interval. Any unrepresented data between any two data points therefore represents a potential source of error in a recording. Overall, the process of digitizing an analog signal will invariably result in a loss of fidelity. The Nyquist theorem states that the sampling rate must be at least twice that of the highest frequency of the signal. Due to the relatively low frequencies sampled on screw EEG this is not an issue with conventional amplifiers, as they can sample >30 kHz. However, failure to take this into account, especially in LFP recordings, will result in a phenomenon known as aliasing in which high frequencies are represented as lower ones, thus compromising the interpretation of the recording. A popular strategy for avoiding this issue is to simply overcompensate the digitizing rate (e.g.. 512 Hz for EEG or 1 KHz for LFP). This can be followed by routine down-sampling techniques found in a variety of signal-processing software to make subsequent analysis more manageable.

Applications

Here we describe the main applications of EEG/LFPs in animals. This list is not exhaustive but represents some of the most common applications in the literature.

Monitoring Vigilance States

Brainwave activity varies as a function of vigilance state and alertness. In animals, vigilance states can be discriminated by recording the electrical field activity with the EEG coupled with LFP recordings in specific arousal-promoting structures in the brain. In addition to EEG, electrodes can be placed on different muscles to record the electromyogram (EMG), as well as around the eyes to record the electro-oculogram (EOG). The combination of these recordings (often called polysomnographic recordings) is used to define states of wakefulness and sleep. In rodents it is possible to distinguish between wakefulness, NREM sleep, and REM sleep. In animal research, polysomnographic recordings using EEG/LFPs are commonly used to study the neural network underlying sleep and wake states and to better understand parasomnias such as sleep apnea, narcolepsy, restless leg syndrome, and insomnia. Similarly, these techniques can also be used to assess the effect of anesthetics and others drugs or to diagnose coma or brain death.

Investigating Epilepsy and Localizing the Origin of Seizures

Using EEG/LFP in experimental animal models for epilepsy has proven to be a successful approach to understanding the underlying mechanisms. An epileptic seizure is caused by massive synchronization of neuronal activity. During a typical seizure, neurons in a restricted area (called the epileptic focus) discharge synchronously, creating a large-amplitude signal in the EEG. Indeed, in epileptic discharges, the membrane potential of neurons changes dramatically and leads to large fluctuations of intra- and extracellular fields that can be readily detected in a clinical environment. Several EEG patterns have been recognized as a signature of epileptiform activity. Absence (petit mal) seizures are brief and characterized by a loss of awareness. During absence seizure, the EEG shows spontaneous spike-wave discharges. Generalized tonic-clonic (grand mal) seizures are usually divided into two phases, tonic and clonic. During the tonic phase the animal abruptly loses consciousness and the skeletal muscles become tense or rigidify. During the clonic phase, the animal's muscles contract and relax rapidly, causing convulsions. During seizure, the initial EEG change may be a low-voltage desynchronization of the EEG. The onset of tonic-clonic seizures is accompanied by rhythmic 15- to 25-Hz activity, decelerating to lower frequencies, resolving to slow waves.

Measurement of Event-Related Potentials and Evoked Potentials in Cognitive Processes

ERPs are voltage fluctuations that can be detected by EEG/LFP in response to the presentation of a stimulus, typically sensory, and are assumed to reflect some form of stimulus processing. ERPs are usually extracted from EEG or LFP recordings (Fishman 2012, Pfurtscheller & Lopes de Silva, 1999) by averaging recording epochs that are time-locked to each of multiple occurrences of sensory, cognitive, or motor events, thereby increasing the signal-to-noise ratio. The ERP pattern depends on the nature of the stimulation, the placement of the recording electrode, and the actual state of the brain. ERPs are used as a clinical diagnostic tool to test the integrity of the sensory pathways in pathological conditions. ERPs are also helpful for studying the time course of cognitive processes such as perception, attention, communication, or memory in normal and pathological conditions, both in humans and rodents.

Neurofeedback

Currently, the most common type of biofeedback is EEG biofeedback, also called neurofeedback. Neurofeedback uses real-time display of EEG to provide a signal that can be used by a person or animal (often a monkey) to receive feedback about brain activity and physiological processes. Biofeedback training is based on operant conditioning in which the participant can search voluntarily for an appropriate strategy to self-regulate subjective sensations and psychomotor activities (Heinrich et al. 2007, Reiner 2008). Physiological processes, such as changes of the brain activity in the expected/desired direction, are reinforced by a positive reward. The development of neurofeedback has led to different clinical applications such as in the treatment of epilepsy, anxiety, substance abuse disorders, insomnia, attention-deficit/hyperactivity disorder, autism, and headaches. There are only a few reports on neurofeedback training in animals. In the 1960s, Sterman and associates were the first to use EEG biofeedback in cats to investigate learned suppression of a previously rewarded cup-press response for food by training them to voluntarily produce a "sensorimotor rhythm," which is an identifiable cluster of rhythms acquired from the sensorimotor cortex during tasks (see review by Sterman 2006). More recently, Phillipens (2010) showed for the first time that nonhuman primates (e.g., marmoset monkey) trained on sensorimotor rhythm activity (α waves) can voluntarily control their brain activity measured by real-time EEG. The development of EEG biofeedback, particularly in nonhuman primates, will definitively help to assess whether significant behavioral changes or improvements in neurophysiological outcome can directly be related to neurofeedback treatment, placebo effects, or other factors.

TABLE 7.1 Necessary Equipment

Equipment	EEG	Comments	LFP	Comments
Subjects	Rodents	Best to use young animals acclimated to their environment	Rodents	Same as EEG
Electrodes	Flathead 1/8-inch-long screws	In our experience, 1/8-inch screws will not puncture tissue. Do not screw in completely. Leaving some threading exposed will help anchor the screws during cementing.	Insulated stainless-steel bipolar wires (125 μm in diameter, <1 mm of tip separation), multicontact laminar probes, silicon/iridium array	Many options are commercially available based on budget and goals. In-house fabrication is an option. EEG screws are sometimes used as ground/reference electrodes.
Wires	Multistrand stainless-steel wire (0.002 inches in diameter)	Solder the tips of these wires to the EEG screws, which will serve as a conductive relay between the screw and the head assembly.	None	Electrodes are either crimped to gold pins or integrated into a board.
Pin/head assembly	Gold-plated pin + custom plug	Gold pins will be soldered to the wire, which will then be inserted into the head assembly. Plugs can be either purchased or fabricated in house.	Multiple options	There are many options based on preference and acquisition hardware used. Commercially available electrodes can be fitted to most commercial headstages.
Amplifier	Multiple options	Many companies offer amplifiers and acquisition systems as a pair. Make sure that the lead coming from the amplifier connects to the head assembly. Decide on the number of EEG channels needed. It is often easier to consult a colleague familiar with these systems. All amplifiers come with built-in analog filters.	Multiple options	Same as EEG
Digital interface hardware	Multiple options	Prices range considerably based on the complexity of the hardware. Systems can control external hardware, can do online/offline analysis, and have onboard filters.	Multiple options	Same as EEG

Conclusion

EEG/LFP offers a relatively noninvasive window into electrical field generation in populations of neurons. With pyramidal cortical neurons fortuitously aligned perpendicular to the axis of the electrodes and the low-pass-filtering properties of the extracellular space, electrodes can not only detect synchronicity (i.e., actively processing neurons) but also localize a signal from broad regions of the brain. Although the use of EEG has decreased somewhat with the introduction of sophisticated anatomical imaging techniques with higher spatial resolution, such as PET and fMRI, the higher temporal resolution and relative ease of implementation of EEG make it likely to be used for many decades to come. The ability of EEG and LFP techniques to bridge the gap between human and animal models of investigation makes them important techniques with great relevance in both clinical and basic science research (Table 7.1).

Frequently Asked Questions

Q: How long are EEG or LFP implants in animals viable?

A: Although there are no standard rules regarding how long these implants can be used, typical experiments are limited to about a month or two due to the fact that, over time, active immunological glial responses to implanted material degrade the tissue surrounding the electrodes. Newer commercially available arrays can last significantly longer through their increased biocompatibility as a result of the manufacturing material used and their flexibility.

Q: What do I do with the waveform data once I've recorded it?

A: Data storage and management can become a burden on computer storage capacity in complex electrophysiological experiments; some EEG/LFP recordings can contain terabytes of content. As a result, using the right signal processing tools and analyses is a topic of active debate within the scientific community. You do not need to be a signal processing expert to conduct EEG/LFP experiments. Most analyses are conducted using only a few mathematical software packages (e.g., MatLab), which has led to countless add-ons, online communities, and toolkits made specifically for electrophysiological processing. Furthermore, acquisition software typically comes complementary with analysis software.

Q: What other techniques can I combine with EEG/LFP recordings?

A: Providing the implantation is successful and there is enough space on the skull, one can very easily implant fiber optics for optogenetic experiments, cannulas for pharmacological infusions, and/or head-fixation posts. Animals can also be acclimated to their headstage and tethers, thus giving researchers the ability to conduct behavioral experiments.

Q: How do I remove noise from my EEG/LFP recordings?

A: Removing noise from an electrophysiology rig can be a constant struggle. As stated above, noise artifacts can be generated from loose contacts between the tether and the animal, myogenic contamination, or simply poor-quality implants. However, 60/50-Hz noise tends to be the most common culprit in disrupting recordings. One safeguard against this source of noise is through the use of a Faraday cage, a metal enclosure around the area in which the animal is placed that acts as a shield from external electrical interference. If noise persists, it will be necessary to investigate aspects of the setup that may not be adequately grounded.

Further Reading

Adrian, E. D., & Matthews, B. H. C. (1934). The Berger rhythm: Potential changes from the occipital lobes in man. *Brain* 57: 355–385.

Buzsaki, G. (2006). *Rhythms of the brain*. New York: Oxford University Press.

Caton, R. (1875). The electric currents of the brain. *Br Med J* 2: 278.

Eccles, J. C. (1951). Interpretation of action potentials evoked in the cerebral cortex. *EEG Clin Neurophysiol* 3: 449–469.

Gibbs, F. A., Davis, H., & Lennox, W. G. (1935). The electroencephalogram in epilepsy and in conditions of impaired consciousness. *Arch Neurol Psychiatry* 34: 1133–1134.

Libenson, M. H. (2009). *Practical approach to electroencephalography*. Saunders, Philadelphia.

Loomis, A. L., Harvey, E. N., & Hobart III, G. A. (1937). Cerebral states during sleep, as studied by human brain potentials. *J Exp Psychol* 21: 127–144.

Nunez, P. L., & Srinivasan, R. (2005). *Electric fields of the brain: The neurophysics of EEG*. New York: Oxford University Press.

Rosenbluth, A., & Cannon, W. B. (1942). Cortical responses to electrical stimulation. *Am J Physiol* 135: 690–741.

Sanei, S., & Chambers, J. A. (2007). *EEG signal processing*. Wiley-Interscience, Chichester.

Schomer, D. L., & Lopes da Silva, F. H. (2011). *Niedermeyer's electroencephalography: Basic principles, clinical applications, and related fields*. Philadelphia: Lippincott Williams and Wilkins.

References

Bédard, C., Kröger, H., & Destexhe, A. (2004). Modeling extracellular field potentials and the frequency-filtering properties of extracellular space. *Biophys J* 86: 1829–1942.

Bédard, C., Kröger, H., & Destexhe, A. (2006). Model of low-pass filtering of local field potentials in brain tissue. *Phys Rev E* 73: 051911.

Berger, H. (1929). Über das Elektrenkephalogramm des Menschen. *Archiv für Psychiatrie und Nervenkrankheiten* 87: 527–570.

Birbaumer, N., Elbert, T., Canavan, A. G., & Rockstroh, B. (1990). Slow potentials of the cerebral cortex and behavior. *Physiol Rev* 70: 1–40.

Buzsaki, G., & Kandel, A. (1998). Somadendritic backpropagation of action potentials in cortical pyramidal cells of the awake rat. *J Neurophysiol* 79: 1597–1591.

Dalal, S. S., Osipova, D., Bertrand, O., & Jerbi, K. (2013). Oscillatory activity of the human cerebellum: The intracranial electrocerebellogram revisited. *Neurosci Biobehav Rev* 37: 585–593.

Destexhe, A., & Bédard C. (2012). Do neurons generate monopolar current sources? *J Neurophysiol* 108: 953–955.

Fishman, Y. I., Steinschneider, M. (2012) Searching for mismatch negativity in primary auditory cortex: deviance detection or stimulus specific adaptation? *J Neurosci* 32: 15747–15758.

Heinrich, H., Gevensleben, H., & Strehl, U. (2007). Annotation: neurofeedback—train your brain to train behaviour. *J Child Psychol Psychiatry* 48: 3–16.

Henze, D. A., Borhegyi, Z., Csicsvari, J., et al. (2000). Intracellular features predicted by extracellular recordings in the hippocampus in vivo. *J Neurophysiol* 84: 390–400.

Jasper, H. H. (1958). Report of the committee on methods of clinical examination in electroencephalography. *EEG Clin Neurophysiol* 10: 370–371.

Jouvet, M. (1969). Biogenic amines and the states of sleep. *Science* 163: 32–41.

Katzner, S., Nauhaus, I., Bunicci, A., et al. (2009). Local origins of field potentials in visual cortex. *Neuron* 61: 35–41.

Lindén, H., Pettersen, K. H., & Einevoll, G. T. (2010). Intrinsic dendritic filtering gives low-pass power spectra of local field potentials. *J Comp Neurosci* 29: 423–444.

Nauhaus, I., Busse, L., Carandini, M., & Ringach, D. L. (2009). Stimulus contrast modulates functional connectivity in visual cortex. *Nature Neurosci* 12: 70–76.

Nelson, M. J., & Pouget, P. (2010). Do electrode properties create a problem in interpreting local field potential recordings? *J Neurophysiol* 103: 2315–2317.

Nunez, P. L., & Srinivasan, R. (2005). *Electric fields of the brain: The neurophysics of EEG*. New York: Oxford University Press.

Pfurtscheller, G., & Lopes de Silva, F.H. (1999). Event-related EEG/MEG synchronization and desynchronization: basic principles, *Clin Neurophysiol*, 110:1842–1857.

Reiner, R. (2008). Integrating a portable biofeedback device into clinical practice for patients with anxiety disorders: results of a pilot study. *Appl Psychophysiol Biofeedback* 33: 55–61.

Riera, J. J., Ogawa, T., Goto, T., et al. (2012). Pitfalls in the dipolar model for the neocortical EEG sources. *J Neurophysiol* 108: 956–975.

Rocher, A. B., Kinson, M. S., & Luebke, J. I. (2008). Significant structural but not physiological changes in cortical neurons of 12-month-old Tg2576 mice. *Neurobiol Dis* 32: 309–318.

Siegel, M., Donner, T. H., & Engel, A. K. (2012) Spectral fingerprints of large-scale neuronal interactions. *Nature Rev Neurosci* 13: 121–134.

Tatum, W. O. (2013). Artifact-related epilepsy. *Neurology* 80: S12–25.

Yuval-Greenberg, S., Tomer, O., Keren, A. S., et al. (2008). Transient induced gamma-band response in EEG as a manifestation of miniature saccades. *Neuron* 58: 429–441.

Voltage-Sensitive Dye Imaging

William Frost and Jian-young Wu

Introduction

Voltage-sensitive dye (VSD) imaging, an optical method of measuring transmembrane potential, has been undergoing constant development since pioneering work published 45 years ago (Cohen et al. 1968, Tasaki et al. 1968). With improving dyes and equipment, the method has gradually become a powerful tool for examining the transmembrane voltage of excitable tissues. Over time, dozens of dyes have been found useful (Grinvald et al. 1982, Gupta et al. 1981, Loew et al. 1992, Shoham et al. 1999), and many analogs have become commercially available. These dyes all have very fast response times (<1 μsec) and excellent linearity over the entire physiological range (Ross et al. 1977), but all show only small fractional changes in absorption or fluorescence. In biological tissues, these fractional changes typically amount to only 10^{-2} to 10^{-5} of the resting light intensity per 100 mV of membrane potential change.

Compared to calcium imaging signals, which have fractional changes of 10–100%, VSD signals are much smaller. While many fast VSDs can provide large fractional changes (up to 30% per 100 mv) in cultured cells (e.g., Miller et al. 2012), fractional changes in tissues are greatly reduced due to background light from nonspecific staining. For this reason, identifying spikes in individual neurons can only be achieved in invertebrate ganglia where neurons are large and sparse (Wu et al. 1993) or in cortex when dye is injected into single neurons (Popovic et al. 2011).

However, high background light and small fractional changes do not necessarily entail a poor signal-to-noise ratio. In this chapter we describe methods for achieving high signal-to-noise ratio recordings of VSD activity in brain slices and invertebrate ganglia. If vibration noise is properly reduced, a signal-to-noise ratio of 5–30 can be achieved in single recording trials, sufficient to identify individual spikes in invertebrate neurons and to examine the initiation and propagation of population neuronal activity in local mammalian circuits in brain slices and *in vivo*. This high sensitivity is critical for examining the dynamics of cortical activity patterns such as spiral waves (Huang et al. 2004, 2010).

This chapter describes the use of a photodiode array for obtaining VSD recordings in both vertebrate and invertebrate preparations. Both of our laboratories use RedShirt Imaging (Decatur, GA) NeuroPDA III systems, built around a WuTech H-469V photodiode array (Fig. 8.1) mounted on the photoport of a conventional compound microscope. The preparation image is focused onto a 19-mm-diameter hexagonal aperture made up of 464 closely packed optical fibers, the other ends of which are each glued to one of the 464 individual photodiodes of the array. A key advantage of a photodiode array system is its suitability for use with absorbance VSDs, most of which, while easily fast enough to record individual action potentials, also yield very small signals. Far beyond current imaging camera chips, the NeuroPDA III can achieve ~21 effective bits of resolution, allowing detection of signals as small as 0.0001% of the resting light level, while sampling the full array at 1,600 frames per second.

The photo-current from each diode is fed into a two-stage amplifier circuit. The first stage converts it into a voltage signal, and the second stage provides 100× amplification. Absorbance dyes must be used with high transmitted resting light levels to maximize the signal-to-noise ratio. Because this bright imaging light gives rise to a several-volt DC offset voltage, on which the 0.05- to 5-mV neuronal optical signals ride, a high-pass hardware filter is inserted before the 100× amplification stage to remove the DC offset, leaving

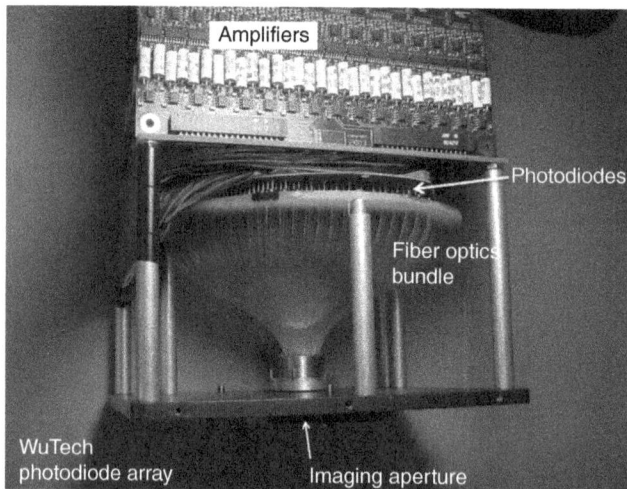

FIGURE 8.1 **Photodiode array**. This device, WuTech 465V, is a commercial version of the recording apparatus developed by the Cohen group (Cohen and Lesher 1986). The WuTech 465V contains 464 silicon PIN photodiodes, each glued onto one optic fiber. All optic fibers form a hexagonal coherent bundle with one end polished as the imaging aperture (see Fig. 8.7A). Each photodiode is wired to a dedicated two-stage amplifier. The first stage converts the photocurrent outflow of the diode into voltage. The second stage subtracts the DC component (resting light intensity) and provides a 100× voltage gain to the AC component (dye signals). This two-stage amplifier circuit significantly increases the effective dynamic range of the system. The integrated design of this diode array reduces both light and electrical interference.

just the neuronal signals. The second-stage amplifier also contains a low-pass filter with a 333-Hz corner frequency, which further improves the quality of the analog signal. The data are then digitized by a 14-bit data acquisition board (Microstar Laboratories, Bellevue, WA) installed in a desktop PC computer. The A/D board samples all 464 optical channels, plus 8 channels of analog data (e.g., stimulus monitors, conventional intracellular or extracellular recordings) at 1,600 frames/sec, fast enough for imaging most signals of interest in both brain slices and invertebrate ganglia. If faster signals are of interest, one can image at higher rates by using defined portions of the full array.

Optical Recording from Vertebrate Brain Slices

Slice Preparation

For imaging neuronal population activity in rodent (rat or mouse) brain slices, the local circuits need to be well preserved, with a large proportion of viable neurons and functioning synapses. Thinner slices usually have a poor signal-to-noise ratio, probably because of more extensive cell damage and poor cell recovery. We prefer slices with a thickness of 400–600 µm, because local circuits can be better preserved. Thicker slices need to be well perfused from both sides with a fast perfusion rate.

Following National Institutes of Health (NIH) guidelines and after approval by the institution's animal use committee, the animals are deeply anesthetized with isofluorane and decapitated. The whole brain is quickly removed and chilled in cold (0–4°C) slicing artificial cerebrospinal fluid (sACSF) for 90 s. The sACSF contains (in millimoles) 208 sucrose, 2.5 KCl, 2 $CaCl_2$, 6 $MgSO_4$, 1.25 NaH_2PO_4, 26 $NaHCO_3$, 10 dextrose. Cortical slices are cut in the sACSF at 0–4°C with a vibratome stage (752M Vibroslice, Campden Instruments, Sarasota, FL) and transferred into a holding chamber containing sACSF at 35°C (bubbled vigorously with 95% O_2/5% CO_2). Brain slices can be cut in several planes to conserve different aspects of cortical circuitry. For example, coronal sections can preserve cortical columns, such as whisker barrels. Oblique sections can preserve thalamocortical connections (Agmon and Connors 1992, MacLean et al. 2006). Slices cut tangentially can best preserve the horizontal connections in layer II–III to observe two-dimensional features such as spiral waves (Huang et al. 2004).

After ~30 min of incubation at 35°C in sACSF, the slices are transferred into a holding chamber with normal ACSF containing (in millimoles) 126 NaCl, 2.5 KCl, 2 $CaCl_2$, 2 $MgSO_4$, 1.25 NaH_2PO_4, 26 $NaHCO_3$, 10 dextrose at 26°C (room temperature). We use a large holding chamber, containing about 500 mL of holding ACSF, which can keep slices viable for 8–12 hours. The holding solution is bubbled with a mixture of 95% O_2 and 5% CO_2 and is slowly circulated by a magnetic stirring bar. Stirring provides a slow but steady convection around the slice for delivering oxygen and washing out noxious molecules produced by the cold shock and the trauma of slicing. Slices must be incubated for at least 2 hours before the experiments to allow for recovery from the acute trauma.

Staining

Brain Slices. Brain slices are stained in the holding chamber with low dye concentrations for 1–2 hours. We use 5–20 μg/mL of an oxonol dye, JPW 1132 (first synthesized by R. Hildesheim and A. Grinvald as RH482; aka NK3630 by Nippon Kankoh-Shikiso Kenkyusho Co., Ltd., Japan, now available from Leslie M. Leow, University of Connecticut at Farmington, les@volt.uchc.edu). The dye is dissolved in a staining chamber containing ~50 mL of ACSF, bubbled slowly with a mixture of 95% O_2 and 5% CO_2 (about 1 bubble/sec) and circulated by a stirring bar. Longer staining times allow the cells deep in the tissue to be evenly stained. Since staining time is long, aeration and convection are critical. Dyes tend to be concentrated on the air–water interface; slow bubbling with large bubbles minimizes this depletion of the dye from the solution. After staining, the slices are transferred back to the holding chamber to rinse out excess dye and for incubation until recording.

Rodent Cortex. For rodent cortex *in vivo* we use the "blue" dyes, RH1691 or RH1938, developed by Amiram Grinvald's group (Shoham et al. 1999). The blue dyes are excited by yellow light (630 nm), which does not overlap with the wavelengths absorbed by hemoglobin (510–590 nm) and thus virtually eliminates heartbeat pulsation artifacts. By comparison, the traditional "red" dyes, such as RH795 or Di-4-ANEPPS, have pulsation artifacts that can sometimes exceed the evoked cortical signals by an order of magnitude (Grinvald and Hildesheim 2004, London et al. 1989, Ma et al. 2004, Orbach et al. 1985, Shoham et al. 1999).

Staining mammalian cortex *in vivo* requires great care with regard to the exposed cortex and the animal's physiological condition. An irritated cortex often leads to poor staining. Briefly, animals are pretreated with atropine sulfate (40 μg/kg intraperitoneally [IP]) approximately 30 min prior to anesthetic induction. The animals are anesthetized with 1.5–2% isoflurane in air via tracheal catheter (16G over-the-needle) and ventilated by a small animal respirator (Harvard Apparatus) at a rate of 60–100/min and volume of 2–4 mL. Rate and volume are adjusted to maintain an inspiratory pressure of ~5 cmH_2O and an end-tidal carbon dioxide value of 26–28 mmHg. A cranial window (~5 mm in diameter) is drilled and bone is carefully separated from the dura, which is left intact. Leaving the dura intact significantly reduces movement artifacts during optical recording (London et al. 1989). Nontraumatic craniotomy is important for optimal staining. Irritated brain can appear reddish (due to increased blood flow), or CSF pressure can increase the potential subdural space, which leads to poor staining. Dexamethasone sulfate (1 mg/kg IP) can be given 24 h prior to the experiment to reduce the inflammatory response of the dura. Voltage-sensitive dye RH 1691 (Optical Imaging) is dissolved at 2 mg/mL in artificial CSF solution, and staining is done through the intact dura mater. Drying dura before staining can increase the dural permeability to the dye. A temporary staining chamber is constructed around the craniotomy with silicone valve grease. ~200 μL of dye solution will stain an area

5 mm in diameter. During staining, the dye solution is continuously circulated by a custom-made perfusion pump. The pump has a battery-powered gear motor that gently presses the rubber nipple of a Pasteur pipette once every few seconds. The tip of this pipette is placed in the staining solution and performs a gentle, back-and-forth circulation of a small amount of dye (~100 μL). Staining lasts for 90 min, followed by washing with dye-free ACSF for >15 min. (For further discussion of staining cortex *in vivo* see Lippert et al. 2007.)

Microscope and Light Source

For imaging brain slices with absorption dyes, an ordinary research microscope with inexpensive low numerical aperture (NA) objectives and a conventional 100-W halogen-tungsten filament lamphouse is adequate to obtain an excellent signal-to-noise ratio (Fig. 8.2a). Köhler illumination is achieved through the microscope condenser (Zeiss, 0.9 NA). For a large imaging field (5 mm in diameter) such as that shown in Figures 8.3, 8.4, and 8.5, we use a 5× 0.12 NA objective (Zeiss). When working with smaller imaging fields, 20× (0.6 NA, water immersion, Zeiss) or 40× (0.75 NA, water immersion, Zeiss) objectives can be used accordingly. When a 5× objective is used, each photodetector of a 464-diode array receives light from a tissue area of 150 μm in diameter.

In our absorption measurements, the resting light intensity is usually about one billion photons per millisecond on each optical detector. The dye molecules absorb most of the light in the wavelength used for the measuring. Unstained tissue slices absorb an insignificant fraction of the light (see Fig. 8.2b).

For *in vivo* imaging of mammalian cortex with fluorescent dyes, high light flux optics are necessary. We use a 5× "macroscope," which provides an imaging field of 4 mm in diameter (Kleinfeld and Delaney 1996, Prechtl et al. 1997). The macroscope was assembled from a commercial video camera lens (Navitar, 25 mm F0.95). This inexpensive camera lens provides a high NA of 0.45, compared to an ordinary 4× microscope objective (~0.12 NA). Such a high NA can gather about 100 times more light, significantly increasing the signal-to-noise ratio for *in vivo* fluorescent imaging. A halogen-tungsten filament lamp (12 V, 100 W, Zeiss) or 630-nm LED (Thorlabs) can be used for illumination.

Optical Filters

The top graph in Figure 8.2b shows the absorption spectrum of a cortical slice stained with dye NK3630 and an unstained slice. The unstained slice has a relatively even transmission with a tendency to absorb less at longer wavelengths (*open circles*). The stained slice has a peak of absorption around 670 nm (*solid circles*) and the peak stays the same after a long wash with dye-free ACSF. The light transmission at 670 nm (peak absorption) through a 400-μm well-stained slice is reduced to 1/10 to 1/50 of that of unstained slices.

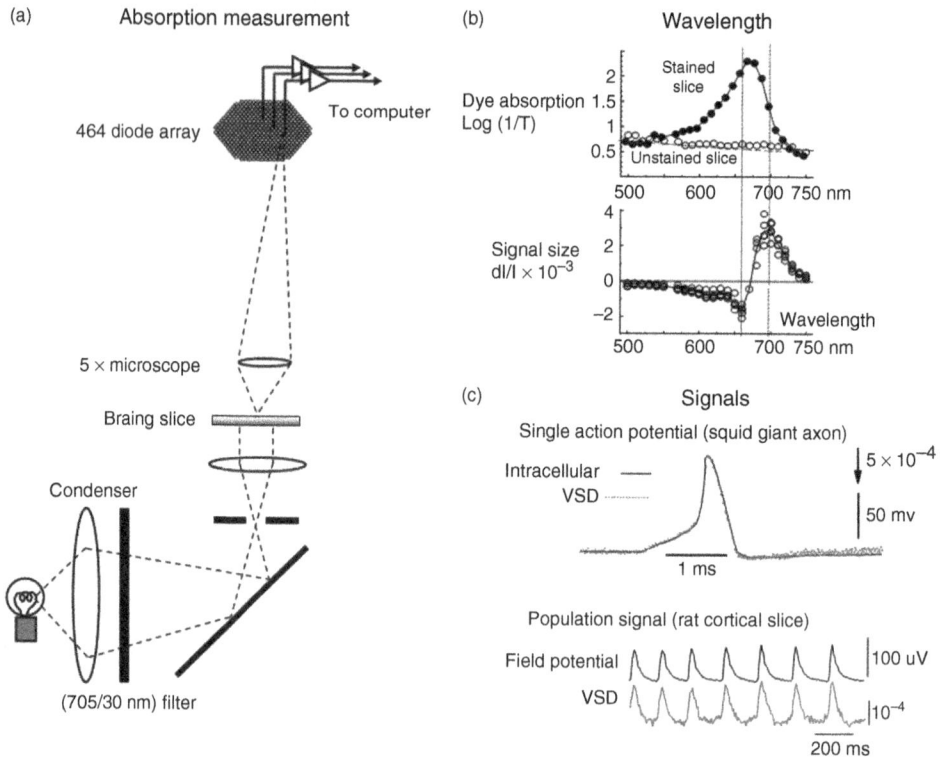

FIGURE 8.2 Absorption signals. a: Rat cortical slices are stained and imaged with a 5x 0.12 NA objective. b, top graph: The absorption spectrum of unstained slices (*unfilled circles*) and a slice stained with NK3630 (*solid dots*). b, bottom graph: VSD signal amplitude at different wavelengths. Note that the maximum amplitude is not at the peak absorbance. (Redrawn from Jin et al. 2002 with permission of Elsevier B.V.) c: Waveform of optical and electrical recordings. Top graph: VSD signal and intracellular recordings during a single action potential in a squid giant axon. The axon was stained with absorption dye XVII and the optical signal was measured at 705 nm (*dotted trace*). The actual membrane potential change (*smooth trace*) was simultaneously recorded by an intracellular electrode. The time course of the VSD signal mirrored precisely the time course of membrane potential change. (Redrawn from Ross et al. 1977, with permission from Springer.) Bottom graph: Population neuronal signals from a rat cortical slice LFP recording (*upper trace*) and a VSD recording (*lower trace*) during a spontaneous 7- to 10-Hz oscillation (Wu et al. 1999). Note that in the rat cortical slice each photodiode receives light from an area of ~330 × 330 μm² of cortical tissue; single action potentials cannot be distinguished.

In the bottom graph in Figure 8.2b, the amplitudes of the dye signal (dI/I) and signal-to-noise ratio are plotted against wavelength. The dI/I and signal-to-noise ratio reaches a maximum at 705 nm. The signal decreases significantly at wavelengths shorter than 690 nm and reaches a minimum at ~675 nm. At wavelengths shorter than 670 nm the signal becomes larger but the polarity of the signal reverses. The signal reaches a second maximum at 660 nm. After the 660-nm peak, the signal decreases gradually and becomes undetectable at wavelengths of 550 nm or shorter. Band-pass filtering around 705 nm should yield the largest signal. Illumination at 675 nm should provide a minimum VSD signal, which can be used for measuring light scattering or for distinguishing VSD from broadband intrinsic signals.

The bandwidth of the filter depends on other factors: In invertebrate ganglia the signal is small and shot noise is a limiting factor. In such cases a wider band-pass allows more light to pass through, thus increasing the signal-to-noise ratio. Alternatively, use of an LED light source eliminates the need for electromechanical shutters and optical band-pass filters (see section on invertebrate ganglion recording). In cortical slice recordings, where signals are large and a long total recording time is needed, a narrow band-pass filter of 705 ± 15 should be used. Any heat filter (infrared filter) in front of the lamp housing absorbs most energy in the 700- to 750-nm range and should be removed from the light path.

For imaging mammalian cortex with the blue dyes, a filter cube (from Chroma or Semrock) is used, with a 630 ± 15-nm excitation filter, a 655-nm dichroic mirror (Chroma Technology), and a 695-nm long-pass filter. Köhler illumination is achieved through the macroscope.

Vibration Noise

Vibration noise is often the dominant noise source when recording absorption signals from brain slices. Vibration causes movement of the image relative to the photodetectors, producing fluctuations in light intensity.

We find that many air tables and active isolation platforms perform poorly in filtering vibrations <5 Hz. A novel isolation stage designed for atomic force microscopy, the "Minus K" table (www.minusk.com), is ~10 times better at attenuating these frequencies than some standard air tables. With the Minus K table, the vibration noise can be reduced to below the level of shot noise at a resting light intensity of ~10^{11} photons/ msec/mm^2.

Data Acquisition and Analysis

NeuroPlex, developed by A. Cohen and C. Bleau of RedShirt Imaging, LLC (Decatur, GA), is a handy program for acquiring and analyzing VSD data. The program can display the data in the form of traces for numerical analysis and pseudo-color images for studying spatiotemporal activity patterns. Data analysis can also be done with scripts written in MatLab (Mathworks).

Several figures in this chapter are examples of data display where the signal from the local field potential (LFP) electrode and that of an optical detector viewing the same location are plotted together. The spatiotemporal patterns of the activity are often presented with pseudo-color maps (Senseman et al. 1990). To compose a pseudo-color map, the signal from each individual detector is normalized to its own maximum amplitude (peak = 1 and baseline/negative peak = 0), and a scale of 16 or 256 colors is linearly assigned to the values between 0 and 1.

The imaging data are acquired at a high rate, usually 1,600 frames per second. Figures for publication usually include only a few frames selected from this large imaging series. The signal-to-noise ratio is usually defined as the amplitude of the VSD signal divided by the root mean square (RMS) value of the baseline noise. Since most population neuronal activity in the neocortex is <100 Hz, low-pass software filters available in NeuroPlex can be used to reduce RMS noise and improve the signal-to-noise ratio.

Sensitivity, Phototoxicity, Bleaching, and Total Recording Time

In cortex, the sensitivity of VSD measurements can be estimated by comparing the measurement with LFP recordings from the same tissue. We have found that the sensitivity of the VSD measurement is comparable to the LFP in both slice (Jin et al. 2002) and *in vivo* (Lippert et al. 2007).

Examples

Over the past 30 years, many authors have published VSD imaging studies of population activity in brain slices (Albowitz and Kuhnt 1993, Bai et al. 2006, Colom and Saggau 1994, Demir et al. 1999, 2000, Grinvald et al. 1982, Hirata and Sawaguchi 2008, Hirota et al. 1995, Huang et al. 2004, Laaris et al. 2000; Mochida et al. 2001, Tanifuji et al. 1994, Tsau et al. 1998, 1999, Wu et al. 1999, 2001). Below, we will show three recent examples of spatiotemporal patterns in rodent neocortex. The signals in these examples must be measured with high sensitivity and in single trials because averaging will obscure the dynamic spatiotemporal patterns.

Spiral Waves

When rodent cortical slices are perfused with carbachol and bicuculline, ~10-Hz oscillations develop in the population (Lukatch and MacIver 1997). Such oscillations can be reliably observed in VSD signals (Fig. 8.3). The spatiotemporal coupling of such oscillations manifests as propagating waves; during each cycle of oscillation a wave of activity propagates through the tissue (Bao and Wu 2003). When cortical slices are sectioned in a tangential plane (Fig. 8.3a), these waves are allowed to propagate in two dimensions with a variety of patterns, such as spirals, target/ring, plane, and irregular waves (Fig. 8.3c). Plane waves have straight traveling paths across the tissue; target/ring waves develop from a center and propagate outward; irregular waves have multiple simultaneous wave fronts with unstable directions and velocities; spiral waves, likely arising out of the interactions of multiple plane waves, appear as a wave front rotating around a center. Each cycle of the rotation is associated with a cycle of the oscillation. Huang et al. (2004) reported observing spiral waves in 48% of their recording trials. Both clockwise and counterclockwise rotations can be observed in the same slice during different oscillation epochs.

In the phase map of a spiral wave, the highest spatial phase gradient is observed at the pivot of the spiral (see Fig. 8.3d, *white dot*), which is defined as a "phase singularity." The presence of a phase singularity distinguishes spiral waves from other types of rotating waves and is the hallmark of a true spiral wave (Ermentrout and Kleinfeld 2001,

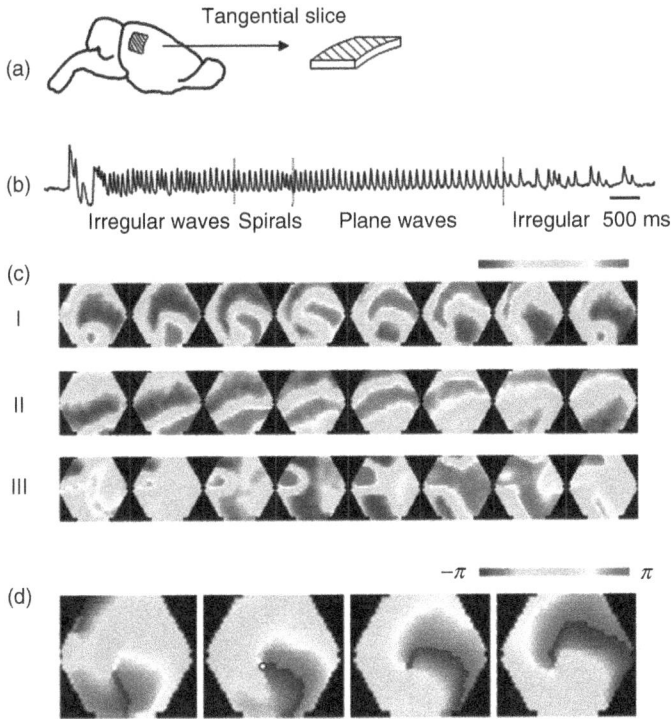

FIGURE 8.3 Propagating waves in rat cortical slice. a: The slice made as a 6 × 6-mm² tangential patch from rat visual cortex. b: VSD signals of oscillations induced by bathing the slice in 100 μM car-bachol and 10 μM bicuculline. Three types of wave patterns during one episode of oscillations (spirals, plane, and irregular) can be identified from imaging data in part c. c: Optical recording images during the oscillation epoch. In each row eight frames (12 ms interframe interval) are selected from thousands of consecutive frames taken at a rate of 0.6 ms/frame. c-I: Spirals, as a rotating wave around a cen-ter. c-II: Plane waves, as a low curvature wave front moving across the field. c-III: Irregular waves, as multiple wave fronts interacting with each other, sending waves with a variety of velocities and direc-tions. d: Phase maps of a spiral wave, depicting the oscillation phase between −π and π according to a linear scale (top right). (Redrawn from data in Huang et al. 2004 with permission of the Society for Neuroscience.)

Jalife 2003, Winfree 2001). Phase singularities in cortical slices have been hypothesized to be associated with small regions containing oscillating neurons with nearly all phases represented between −π and π (Huang et al. 2004), with the phase mixing resulting in amplitude reduction in the optical signal. In Figure 8.4 each detector covered a circular area of 128 μm in diameter (total field of view 3.2 mm in diameter). Signals from all detectors showed high-amplitude oscillations before the formation of spirals (see Fig. 8.4a), traces A–E before the first broken vertical line). During spiral waves, the phase singularity slowly drifted across the tissue. The four detectors, A–D, alternately recorded reduced amplitude as the spiral center approached each detector in turn. Such ampli-tude reduction was localized at the spiral center, and this reduced amplitude propa-gated with drift of the spiral center (see Fig. 8.4a, traces A–D). At locations distant from the spiral center (e.g., location E in Fig. 8.4a), the amplitude remained high during all

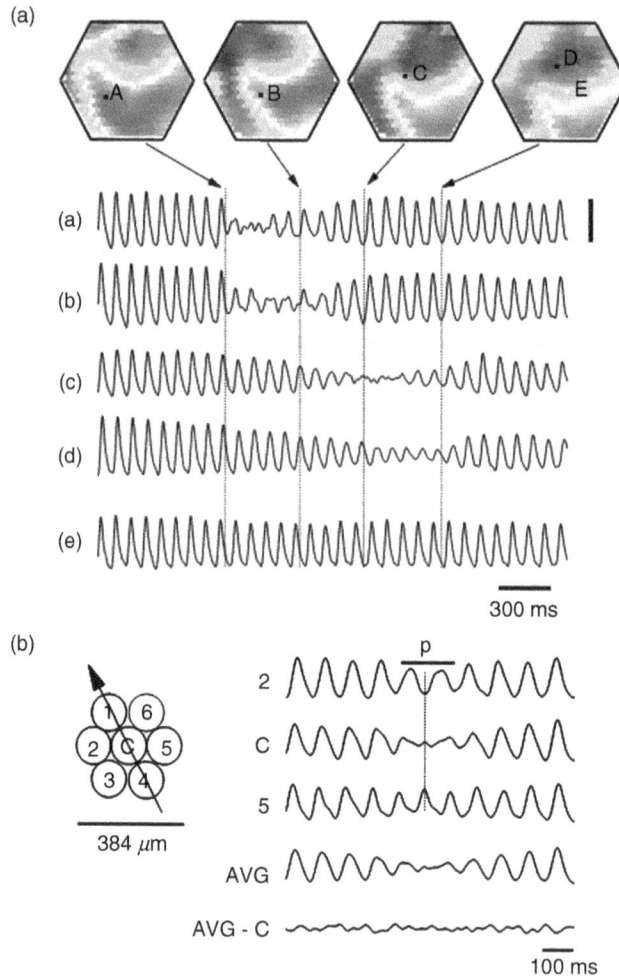

FIGURE 8.4 **Amplitude reduction at spiral singularity**. a: Representation of signal amplitude in four consecutive frames. The spirals started at about the time of the first image on the left and ended at the last image. Traces A–E: Optical signals from five detectors indicated in individual frames. The *broken vertical lines* mark the times the images were taken. During spiral waves the four detectors, A–D, sequentially recorded a reduced amplitude as the spiral center approached each detector. Detector E recorded a high-amplitude signal throughout the entire period because the spiral center never reached that location. Calibration: 2×10^{-4} of resting light intensity. B, left: The trajectory (*arrow*) of a spiral center moving through a group of optical detectors (1–6 and C). B, right: Signals from the detectors mapped at left. *Black bar* (p) indicates the time when the spiral hovered over the center detector. The surrounding detectors (2, 5) did not show amplitude reduction but the oscillation phases exactly opposed each other. AVG: Averaged signal of six surrounding detectors shows an amplitude reduction similar to that on the center detector C. AVG - C, numerical subtraction of trace C and AVG. These results indicate that the spiral center was fully confined within the area of the central detector. They also suggest that the amplitude reduction on trace C is caused by a superimposition of multiple phases surrounding the center detector. (Adapted from Huang et al. 2004 with permission of the Society for Neuroscience.)

rotations of the spiral. In plane or ring waves, no localized region of oscillatory amplitude reduction was seen.

To further confirm that amplitude reduction was caused by superposition of antiphased oscillations, we examined signals surrounding a spiral center. Figure 8.4b shows the signals from a group of detectors when a spiral center drifted over the center detector (C). As the spiral center hovered briefly (~200 ms, two rotations, marked as p above the traces in Fig. 8.4b) over the center detector, the amplitude was reduced (trace C). Simultaneously, the six surrounding detectors (1–6) did not show amplitude reduction, but the oscillation phases exactly opposed each other symmetrically across the center (traces 2 and 5 are shown). These results suggest that the area with amplitude reduction was less than or equal to the size of our optical detector's field of view (128 μm in diameter). When we added the signals from the six surrounding detectors, the averaged waveform showed a similar amplitude reduction as the center detector (see Fig. 8.4b, trace AVG). This combined signal was nearly identical to the central signal, as demonstrated by the small residual when the two signals were subtracted (see Fig. 8.4b, trace AVG - C). These findings strongly support the hypothesis that amplitude reduction is not caused by inactivity of the neurons, but rather by the superimposition of multiple widely distributed phases surrounding the singularity, and that the spiral center was fully confined within an area of ~100 μm in diameter.

This example also demonstrates that high sensitivity and single trial recordings are essential for studying dynamic events such as spirals. At locations near the spiral singularity, for example, the VSD signal is reduced 10-fold and requires an order of magnitude better sensitivity than if one were observing simple oscillatory events without such dynamic spatiotemporal patterns.

Locally Coupled Oscillations

Fast VSD imaging has the appropriate spatiotemporal resolution to reveal multiple local oscillations in brain slices. Figure 8.5 shows an evoked network oscillation (~25 Hz) in neocortical slices, trigged by a single electrical shock (Bai et al. 2006). In a train of oscillations, for different cycles the waves may be initiated at different locations, suggesting that local oscillators are competing for the pacemaker to initiate each oscillation cycle. The first spike was initiated at the location of the stimulating electrode (see Fig. 8.5, *black cross* in first top-row image) and propagated across the slice. Two oscillation cycles, 3 and 7, were initiated from two different locations (see Fig. 8.5, *black cross* in middle row and upper black cross in bottom-row images) and propagated in concentric patterns surrounding their respective initiation foci. This pattern indicates that oscillations are not synchronized over space; multiple pacemakers exist and may lead different cycles alternately in turn. On the other hand, local oscillators are not completely independent, because there is a propagating wave accompanying each oscillation cycle. If the oscillators were completely independent, there would be no propagating waves. VSD imaging provides a powerful tool to study such spatial coupling of local oscillators in neocortical circuits.

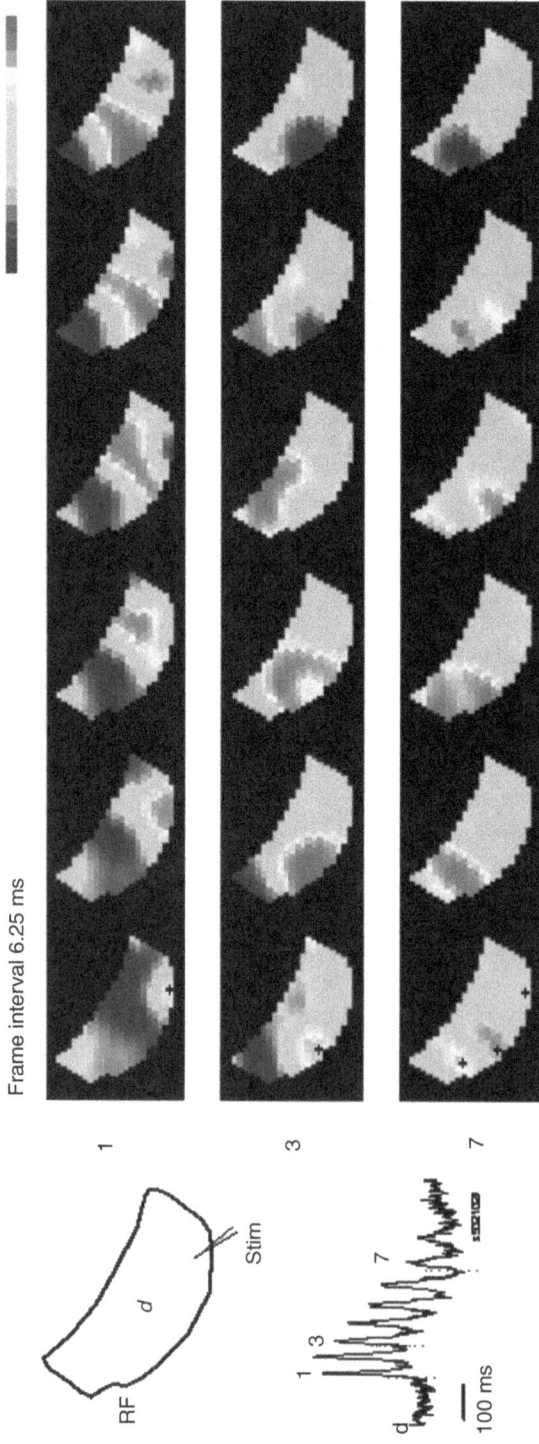

Frame interval 6.25 ms

FIGURE 8.5 Initiation foci of an evoked cortical oscillation. Left: Orientation of the slice and stimulation sites (top). The VSD signal from one optical detector *d* (bottom). Right: Images of three oscillation cycles (1, 3, 7). All images are snapshots of 0.625 ms; the interval between images is 6.25 ms. Each row of images starts at the time marked by a *vertical broken line* in the trace display in the left panel. The initiation site was determined as the detector with the earliest onset time and marked with *black crosses* in the first image of each row. The first wave (1) initiated at the location of the stimulating electrode and propagated as a wave to the lateral (top row). Two following oscillation cycles (3 and 7) initiated from two different locations. All three initiation foci are marked at the first image of the bottom row for comparison. This optical recording trial contained 1,100 images (~0.7 sec); only a few selected frames are shown for clarity. (Adapted from Bai et al. 2006, with permission of the American Physiological Society.)

High-Sensitivity Optical Recordings in the Cortex *in Vivo*

When the cortex is properly stained, the sensitivity of VSD recording can be comparable to that of LFP recordings (Fig. 8.6). Since the signal is transmembrane potential, it is not affected by current source/sink pairs as it is in LFP recordings; therefore, high spatial resolution can be achieved (e.g., 464 recording sites in an area of ~5 mm in diameter). In this experiment, cortical activity was recorded with VSD imaging and LFP (electro-encephalogram in Fig. 8.6) simultaneously from the same tissue. Comparing these two signals, we found that most of the peaks in the LFP showed a corresponding event in the VSD trace. Under 1.5% isoflurane anesthesia, infrequent bursts of spontaneous activity were recorded in both electrical and optical recordings and most events within each burst correlated well between LFP and VSD signals (see Fig. 8.6b, top). Adjusting the level of anesthesia allowed us to further verify the correlation between the LFP and VSD signals. When the level of anesthesia was lowered, both VSD and LFP signals showed continuous spontaneous fluctuations (see Fig. 8.6b, bottom). The fluctuations in the VSD signals are unlikely to be baseline noise, since the baseline noise is much smaller in the quiescent segments of Figure 8.6b, which were obtained from the same location in the same animal. Instead, these fluctuations are likely to be biological signals from local cortical activity, probably sleep-like oscillations or up/down states (Petersen, Hahn, et al. 2003).

LFP and VSD signals were frequently disproportionate in amplitude. In Figure 8.6b, events labeled with *dots* had a higher amplitude in the VSD trace, while events labeled with *triangles* had a larger amplitude in the LFP trace. Events labeled with *diamonds* were seen in the LFP but not in the VSD signals. Under lower anesthesia, while the correlation between the LFP and VSD recordings decreased, many events in the LFP trace were also seen in the VSD traces (exceptions are marked with *squares* in Fig. 8.6b). VSD recordings from two adjacent locations also show good correlation (bottom two traces in each panel of Fig. 8.6b), further indicating that the fluctuations in the VSD traces are not noise. Since almost every event in the LFP can be recorded clearly in VSD imaging, the sensitivity of VSD measurement is comparable to that of LFP recordings. However, VSD and LFP signals have a low overall correlation of ~0.2 between 3 to 30 Hz (which includes 90% of the power). This low overall correlation is probably due to some LFP peaks originating from deep cortical layers or subcortical sources.

The correlation between optical traces is much higher and each peak in the signal appears at many locations (Fig. 8.6b, bottom traces). The high correlation between optical traces is not due to light scattering or optical blurring between channels, because there is a small timing difference between locations, resulting in propagating waves in spatiotemporal patterns (Brown et al. 2009, Ferezou et al. 2006, Gao et al. 2012, Huang et al. 2010, Lippert et al. 2007, Petersen 2007, Petersen, Grinvald, and Sakmann 2003, Petersen, Hahn, et al. 2003, Slovin et al. 2002, Xu et al. 2007). At a light intensity that

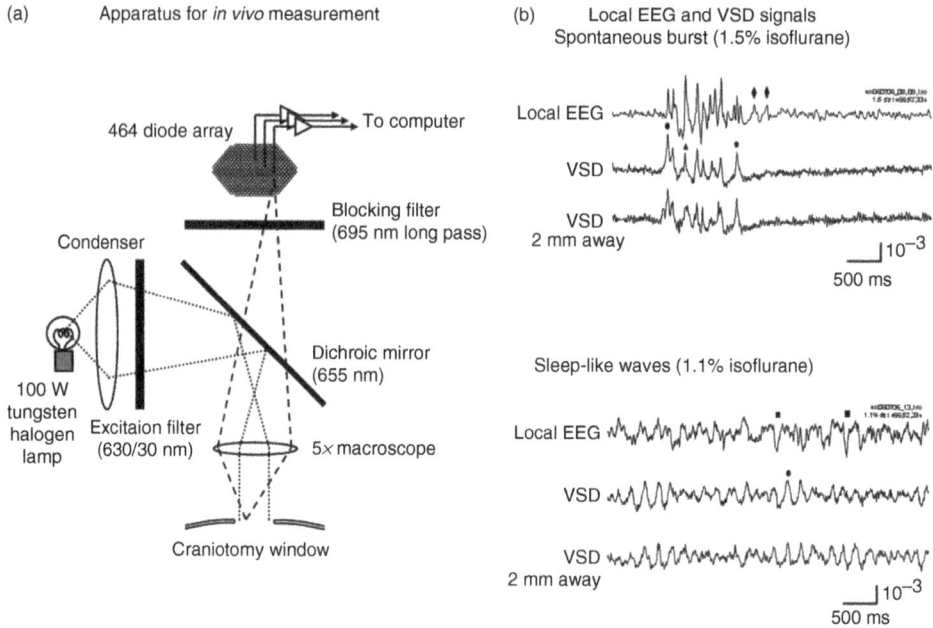

FIGURE 8.6 **Fluorescent signals from rat visual cortex**, *in vivo*. a: Apparatus for *in vivo* imaging of mammalian neocortex. A 5× "macroscope" is used to provide high NA (Kleinfeld and Delaney 1996, Prechtl et al. 1997). The macroscope was assembled from a commercial video camera lens (Navitar, 25 mm F0.95). This inexpensive camera lens provides a high NA (0.45), which gathers ~100 times more light than an ordinary 5× microscope objective. The macroscope projects an imaging field of 4 mm in diameter to the 19-mm diode array aperture, equivalent to a 5× optical magnification. Halogen-tungsten filament lamp (12 V, 100 W, Zeiss) or 630-nm LED (Thorlabs) can be used for illumination. b: Local EEG and VSD signals. The top two traces in each panel are LFP and VSD recordings from the same location in the visual cortex of rat. The bottom traces in each panel are VSD recordings taken 2 mm away from this location. A correlation plot of the VSD and LFP of the raw data is presented on the right. Top traces: Under 1.5% isoflurane anesthesia, spontaneous spindle-like bursts were recorded. *Dots* mark events in which VSD has a higher amplitude than LFP. *Triangles* mark events in which the LFP has a higher amplitude than the VSD recording. *Diamonds* mark events seen in the LFP but not in the VSD recording. *Bottom traces:* Recordings from the same animal about 5 min later, with isoflurane anesthesia lowered to 1.1%. Most of the peaks in the LFP are also seen in the VSD recording, but the correlation between the two signals is lower. The baseline fluctuations cannot be attributed to noise, since recordings from the same preparation had almost no spontaneous fluctuations when the anesthesia concentration was elevated. (Modified from Lippert et al. 2007 with permission of American Society of Physiology.)

results in the signal-to-noise ratio shown in Figure 8.6, we can record up to 80–100 trials of 10 s each, which is sufficient for many types of experiments. With longer recording times, the signal amplitude decreases and epileptiform spikes occasionally develop, probably due to phototoxicity.

Optical Recording from Invertebrate Ganglia

Due to their large neurons and relatively simple nervous systems, invertebrates have long been used to study how networks of specific, identifiable neurons process sensory input, generate behavioral motor programs, and store memory (Clarac and Pearlstein 2007, Sattelle and Buckingham 2006). For many years invertebrate studies have been limited to recording from two to four neurons at a time using sharp electrodes. However, recording with VSDs has steadily been improving as a method for recording large numbers of neurons simultaneously (Briggman and Kristan 2006, Cohen et al. 1989, Frost et al. 2010, Hill et al. 2014, London et al. 1987, Wu et al. 1988, Zecevic et al. 1989). Here we describe our use of fast absorption VSDs to record action potentials from ~100 neurons simultaneously in suitable invertebrate preparations.

Recording Apparatus

We obtain our *Tritonia diomedea* ganglion recordings using the same RedShirt Imaging photodiode array described above for vertebrate studies, mounted on an Olympus BX series upright microscope. The ganglion image is focused on a closely packed hexagonal grid of 464 optical fibers (Fig. 8.7A, B), which bring the light to the 464 photodiodes of the array (see Fig. 8.1). This design ensures that each photodiode sees a different area of the neuronal population. (See Frost et al. 2010 for a more detailed description of methods.)

In our invertebrate imaging we use Olympus 10×0.6 NA and 20×0.95 NA water immersion objectives. The high NA of these objectives allows us to use a lower lamp brightness to obtain the high transmission light levels needed to maximize the signal-to-noise ratio, minimize dye bleaching, and allow longer imaging sessions (e.g., several 2-min acquisitions per preparation). We currently use a high-power 735 ± 20-nm LED light source (Thorlabs M735L2-C1) to transilluminate the preparation, driven with a DC power supply (Thorlabs DC2100). This stimulator-controlled illumination method eliminates the need for a band-pass optical filter and an electromechanical shutter, the latter of which can introduce unwanted vibrations that add artifact signals to the imaging data. Köhler illumination is achieved through the condenser of the BX microscope. Because fast absorption VSDs typically yield small signals (action potentials typically produce a 0.01–0.1% change in the resting light level), any vibration of the image produces movement of contrast edges on the diodes, resulting in artifacts that contaminate the neuronal signals. We further minimize vibration by mounting the microscope on a Minus K MK26 spring-based vibration isolation workstation, and by imaging with the above water immersion objectives.

The *Tritonia* brain, consisting of the cerebral, pleural, and pedal ganglia, is removed from the animal and pinned to the Sylgard-lined coverslip bottom of the recording chamber (Siskiyou PC-H). The preparation is then placed under the microscope and perfused at physiological temperature with the VSD RH155 (Anaspec) for 30–60 min. Next, the curved upper surface of the sheathed ganglion is flattened using a coverslip fragment held in place by small mounds of Mack's earplug silicone putty positioned on the chamber floor. This procedure increases the number of in-focus neurons, and hence the number detected in the imaging.

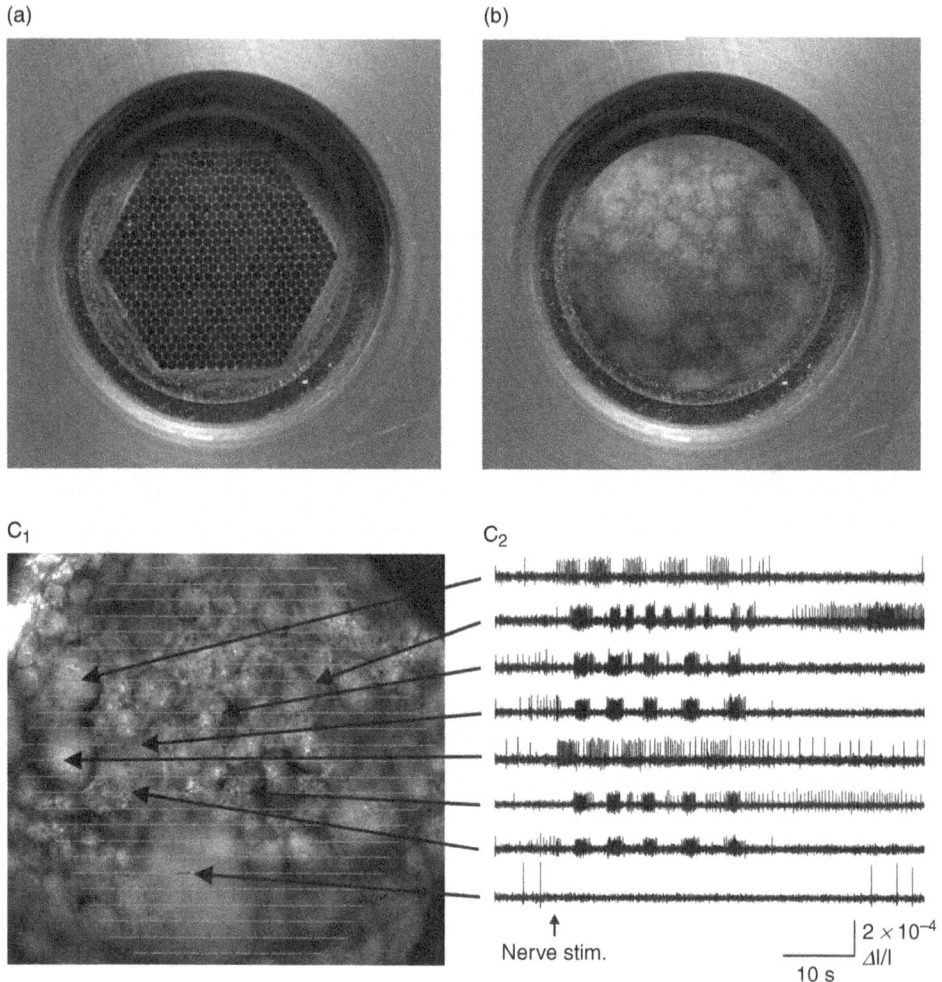

FIGURE 8.7 How raw data are acquired and displayed by the photodiode array. A: The hexagonal region is the port of the photodiode array, composed of the cut ends of 464 fiber optics that each conveys light to a different photodiode (see Fig. 7.1 for additional views). B: Depiction of the image of the VSD-stained ganglion, focused on the array port, to show how each diode sees the light coming from a different region of the neuronal population. Larger neurons are seen by several diodes, and some diodes see more than one neuron. These redundant and mixed signals are processed by independent component analysis into single neuron traces. C: Individual diode recordings of the *Tritonia* rhythmic swim motor program, imaged from the dorsal pedal ganglion. C_1: Live-linked array data superimposed over an aligned picture of the imaged ganglion, as they appear in NeuroPlex immediately after data acquisition. C_2: Clicking on the diode traces in C1 displays the software-filtered raw data collected from each ganglion location. *Arrows* indicate the ganglion location for each action potential trace. Several known neuron types are seen, including neurons that burst in the dorsal phase of the swim motor program, others that burst in the ventral phase, and an additional class that bursts twice per cycle.

New Neurons Are Readily Identified Using the Raw Data Alone

Immediately upon acquisition, the optical data are displayed as 464 miniature traces, positioned at their respective array positions, live-linked to the data (see Fig. 8.7C_1). We routinely superimpose a photograph of the imaged ganglion with the data, using a digital camera mounted on one arm of a dual photoport. One can then simply select different regions of the ganglion to localize neurons that respond to a given stimulus or that fire rhythmically during known motor programs (see Fig. 8.7C$_2$). In this way, new neurons of interest can be identified immediately upon data acquisition. Furthermore, by focusing at different depths, neurons at different levels of the ganglion can be selectively recorded.

Independent Component Analysis Spike Sorting to Obtain Single Neuron Traces

More sophisticated questions, such as whether a given network changes in size with learning or whether the recorded neurons fire in different ensembles at different times, require further processing of the data. In our datasets, many diodes record action potentials from multiple neurons, and many of the larger neurons are recorded redundantly by several diodes, making it impossible to generate single neuron traces, and thus know how many unique neurons were recorded, from the raw data alone. A further complication is that diodes often have movement- or vibration-related artifacts mixed into the data, due to the fact that any movement of a contrast edge results in a change in transmission brightness on the diodes. To overcome these problems, we routinely use independent component analysis (ICA) to spike-sort the raw data, which returns a single, artifact-free trace for each individual neuron (Fig. 8.8; artifacts are also recognized as independent components and are relegated to their own traces). An additional advantage is that the resulting spike-sorted traces also have a better signal-to-noise ratio than the raw data traces. The accuracy of this rapid and objective blind source separation approach to spike sorting was recently validated using simultaneous intracellular recordings with sharp electrodes in *Tritonia* and *Aplysia* (Hill et al. 2010). The MATLAB-based spike-sorting software we use runs on any Windows PC and is available at http://cnl.salk.edu/Research/ICA/Optical/. Two useful computer features are significant RAM, which determines the upper duration limit of the data files that can be processed, and a fast processor, which determines the time taken for the spike sorting. A commercial 64-bit Windows laptop with a 2.5-Gz processor and 4 GB of RAM can spike-sort a 2-min data file collected at 1,600 Hz into independent components in approximately 15 min. The analysis also provides the XY ganglion location of each recorded neuron.

Advantages of Certain Preparations

We have obtained excellent optical recordings of action potential traces from >100 neurons, from *Aplysia californica* and *Tritonia diomedea*. These opisthobranch mollusks may be particularly well suited for such recordings due to their large neuronal somata (most are 50–200 μm in diameter) and to the fact that the somata display overshooting action potentials. In addition, the somatic plasma membranes of these and related species are highly involuted (Longley 1984; and our unpublished data). These infoldings have been estimated to increase the somatic surface area by 7-fold (Mirolli and Talbott 1972), which would increase the effective dye density per soma and thus the strength

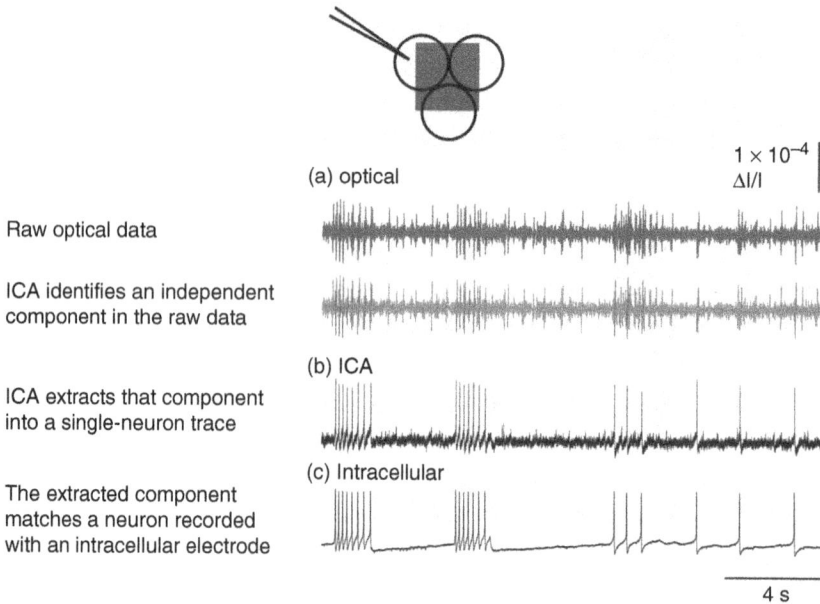

FIGURE 8.8 Spike sorting by ICA. In invertebrate ganglia, each photodiode may record signals from multiple neurons, and many photodiodes may redundantly record signals from the same neuron. ICA performs blind source separation to extract each neuron's action potentials into a separate trace. A, top trace: Optical data from a single photodiode (in diagram and traces), which recorded action potentials originating from multiple neurons. A, bottom trace: By comparing activity in all 464 photodiodes, ICA identifies the action potentials originating from a particular neuron. B: ICA extracts the information common to that neuron from all raw data traces and places it into a new trace as an independent component. C: In this experiment, a sharp electrode placed into that neuron confirms the accuracy of the ICA spike sorting.

of the optical signals. We have also obtained high-quality action potential signals in the marine mollusk *Pleurobranchaea* but failed to in the crab stomatogastric ganglion (unpublished data), where neurons lack overshooting action potentials.

Large Numbers of Neurons Can Be Simultaneously Recorded

In our laboratory we have focused on rhythmic motor networks for known behaviors in the marine mollusks *Tritonia* and *Aplysia*. Rhythmic motor programs have the advantage that the participating neurons are readily apparent in the optical recordings and their behavioral relevance is known. For example, Figure 8.9 shows optical recordings of neurons in the cerebral and pedal ganglia of *Tritonia* during the animal's stimulus-elicited escape swim motor program. Here both well-known and previously undescribed neurons appear in the recordings. Figure 8.9A shows a sharp-electrode recording from three well-studied neurons of the *Tritonia* swim central pattern generator. Figure 8.9B shows an optical recording from the same dorsal cerebral-pleural ganglion region, which detected the firing of all five known swim central pattern generator neurons located on that side (C2, DSI 1–3, and VSI-A). Previously undescribed rhythmic cerebral ganglion neurons are also evident in the recording (adapted from Hill et al. 2012). Figure 8.9C shows an optical

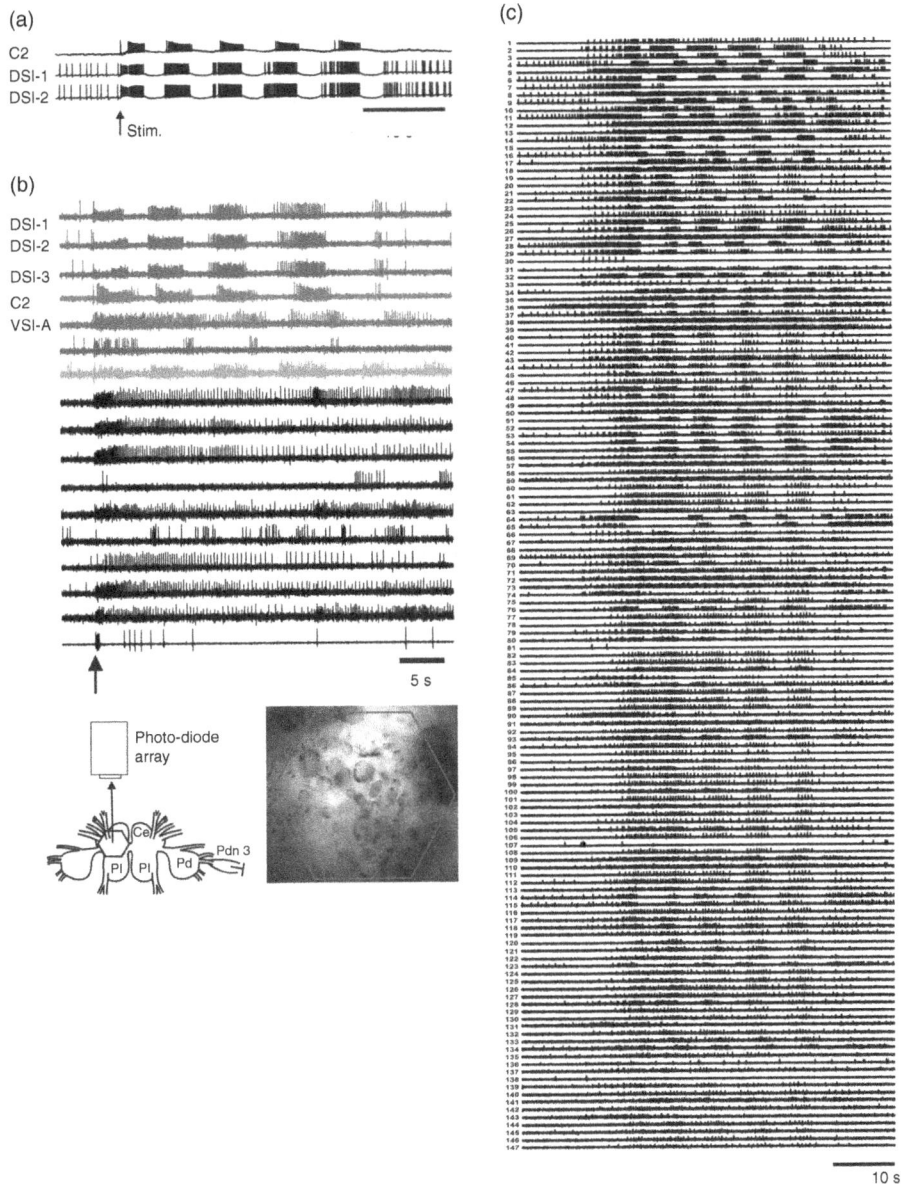

FIGURE 8.9 Optical recording allows large numbers of neurons to be recorded simultaneously. A: Traditional use of sharp electrodes to study the *Tritonia* swim network, elicited via a brief 10-Hz stimulus to pedal nerve 3. Shown are three well-known members of the swim central pattern generator, located in the dorsal cerebral ganglion. Each burst of action potentials is associated with one dorsal or ventral body flexion in the intact animal. Cal. = 10s. B: Identification of known and new neurons in the optical recordings. This recording was of the region of dorsal cerebral ganglion containing most of the known *Tritonia* swim central pattern generator neurons. All known pattern generator neurons on the dorsal side of the brain, including the three DSIs, C2, and VSI-A, were picked up in the optical recordings, readily identified here by their locations and firing patterns during the swim motor program. Previously undescribed rhythmic neurons were also recorded, indicating how easily new neurons of interest can be identified using this approach. (Modified from Hill et al. 2012.) C: ICA spike-sorted optical recording of 147 dorsal pedal ganglion neurons firing during the swim motor program. The eliciting stimulus was seven pulses to PdN3, delivered at 1 Hz.

recording of 147 neurons in the dorsal pedal ganglion during a stimulus-elicited swim motor program in *Tritonia*. Neurons previously described from sharp-electrode studies, such as the DFN-A, DFN-B, VFN, and Type III flexion neuron groups (Hume et al. 1982), which fire reliably on all cycles of the motor program, were recorded in abundance.

New Neurons Can Be Identified and Penetrated Immediately, Accelerating the Study of Neural Circuits

Having the XY ganglion locations of all optically recorded neurons immediately available during an imaging experiment makes it possible to seek them in the same preparation with sharp electrodes to begin studying their properties and connectivity. To facilitate such experiments, we designed and constructed a novel hybrid imaging stereomicroscope, which has switchable objective lenses, one a high-NA water immersion lens for optical recording and the other a standard stereomicroscope objective for sharp-electrode recording (Frost et al. 2007). Figure 8.10 shows an experiment using this microscope where, in just two preparations, two previously undescribed neuron groups participating in the *Tritonia* swim motor program were identified and characterized. The ventral side of the cerebral and pleural ganglia have been little explored by sharp-electrode studies. In our very first two ventral imaging preparations we noted the presence of a previously unknown single neuron in the cerebral ganglion (see Fig. 8.10A1,2) and a bilateral cluster in the pleural ganglion (see Fig. 8.10B1,2) that burst in phase with the swim motor program. In both cases we then switched to the stereo objective and penetrated these same neurons with sharp electrodes (see Fig. 8.10A3, B2). In one of the preparations we penetrated both neuron types and found them to fire in antiphase with one another during the motor program (see Fig. 8.10C). In a subsequent sharp-electrode-only experiment we found the ventral cluster neurons to be excited by central pattern generator neuron VSI-B (see Fig. 8.10D), and dye fills (not shown) showed the cluster neurons to send processes out cerebral nerves to the periphery, consistent with their being a new class of swim flexion efferent neuron. The experiments identifying these two new neuron groups were conducted in two preparations in a single day, illustrating the potential of large-scale optical recording for accelerating the study of neural circuits in both well-established and previously unexplored invertebrate preparations.

Optical Recordings Reveal Previously Undescribed, Variably Participating Neurons

Published sharp-electrode studies have typically depicted members of rhythmic neuronal networks as firing faithfully on every cycle of their motor programs. Given this perspective, we were surprised to find that while the majority of neurons in our optical recordings did fire in this expected way, a significant number participated in a surprisingly casual manner. In *Tritonia* recordings, for example, several neurons in any

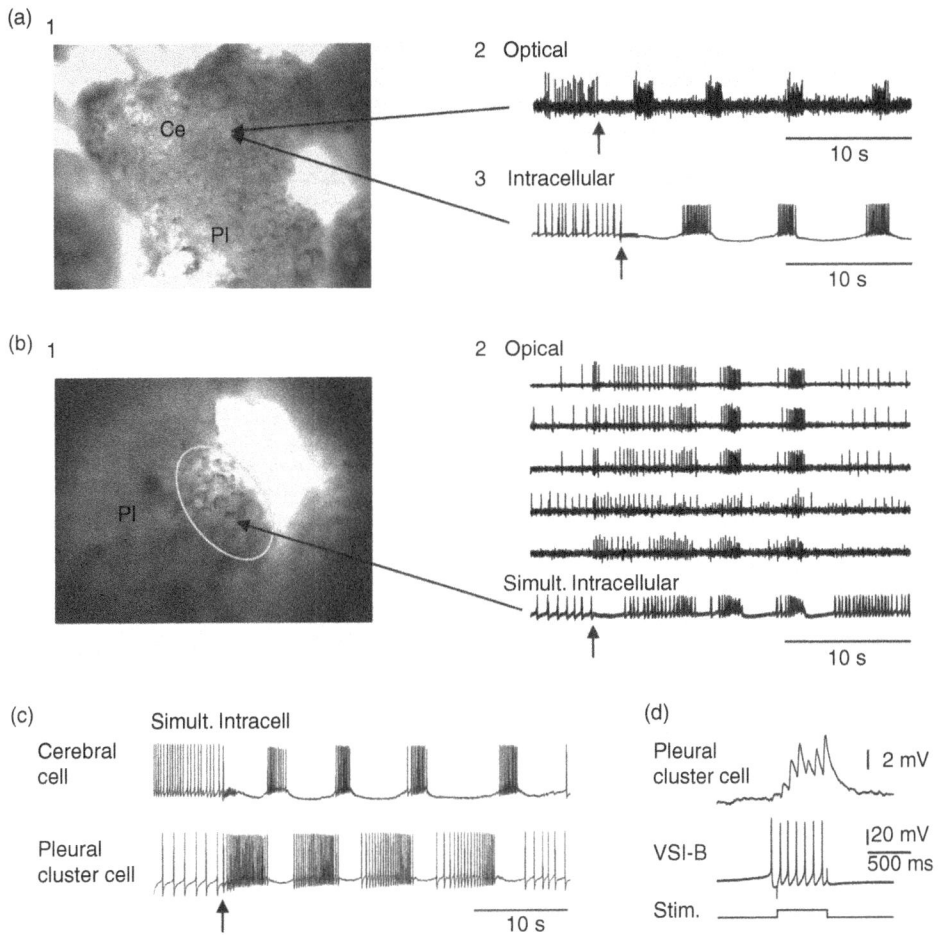

FIGURE 8.10 Optical recording readily identifies new neurons that can be penetrated with sharp electrodes in the same preparation. A: 1 and 2 indicate location and initial optical recording of a new ventral cerebral ganglion neuron in *Tritonia*, firing during a nerve stimulus elicited swim motor program. 3: The same neuron was later found and recorded with a sharp electrode in the same preparation. The two diagonal arrows indicate the neuron. B: 1 and 2 indicate location and initial optical recording from a new bilateral cluster of ventral pleural ganglion neurons, firing during a nerve-elicited swim motor program. The optical traces come from neurons in the encircled area. During this recording, one neuron from the cluster was simultaneously recorded with a sharp electrode. C: Simultaneous sharp-electrode recordings of the two new neuron types, from the same preparation in which the cerebral neuron was discovered, revealing that they fire in antiphase with one another during the swim motor program. D: Neuron VSI-B of the swim central pattern generator excites the pleural cluster cells. "Stim" indicates the depolarizing constant current pulse injected into VSI-B. Upward arrows in all panels indicate onset of the brief PdN3 stimulus used to elicit the swim motor programs. The optical traces in A2 and B2 were obtained after ICA spike sorting. *Ce* and *Pl* stand for the cerebral and pleural ganglia, which are fused in the animal.

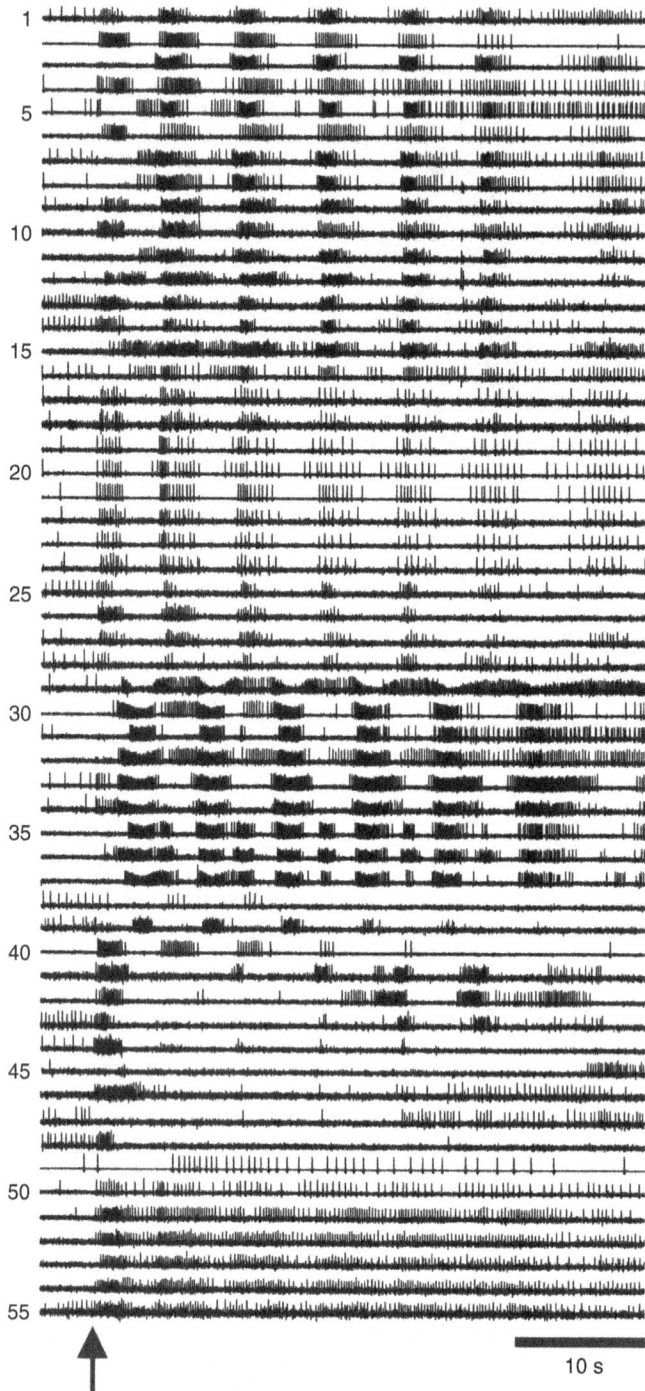

FIGURE 8.11 Optical recording reveals the unexpected presence of variably participating neurons. Recording of 55 neurons during the *Tritonia* swim motor program. While most neurons fire reliably on each cycle, several neurons can be seen to join the motor program late, leave it early, or skip internal cycles. Such variably participating neurons were immediately apparent in the optical recordings and were subsequently found in several sharp-electrode recordings in dye-free saline. (Modified from Hill et al. 2012.)

given cerebral or pedal ganglion recording are typically found to join the motor program late, leave it early, skip middle cycles, or fire on just one or two internal cycles (Fig. 8.11). Neurons behaving this way were subsequently observed in sharp-electrode recordings of motor programs in normal, dye-free saline, indicating that they represent a newly recognized class of neuron that participates variably, cycle by cycle, in the swim rhythm (Hill et al. 2012). Such variably participating neurons were also observed in optical recordings of the *Aplysia* escape locomotion rhythm. This finding reveals that networks can be composed of shifting coalitions of neurons, with individual participants wandering in and out of the network while the motor program is running. This view is consistent with a notion of neural networks having functional wiring diagrams that are highly plastic, not just episode to episode, as in learning, for example, but also moment to moment as they run.

Conclusion

Imaging with VSDs provides information about when and where network activity occurs and how it spreads. In cortical tissue, stationary oscillations recorded with a single LFP electrode appear as propagating waves in the VSD imaging. The waves can have complex patterns such as spirals or multiple wave fronts. Such fascinating large-scale features cannot be discerned by recording with conventional methods at any one site. Cortical tissue is rich in neuropil, with a relatively large area of stained excitable membrane per unit volume. Because during population events cortical neurons are usually activated in ensembles, the combined dye signals can be very large. In optimal conditions 464 locations in ~1 mm^2 of cortical tissue can be monitored simultaneously with sensitivity better than or comparable to that of LFP electrodes, which record activity at a single site. These factors make VSD imaging a useful tool for analyzing spatiotemporal dynamics in cortical local circuits as well as neuronal interactions during information processing.

Beyond the approaches described here, genetically engineered dyes are emerging that can be targeted to specific neuron classes, allowing the contributions of different neuron types to cortical population signals to be assessed.

Studies of invertebrate networks have long been hindered by the limited number of individual neurons, typically two to four, that can be simultaneously recorded using conventional electrodes. Not only does this slow the process of working out basic network structure, but such a restricted view of network activity greatly hinders exploration of many issues of scientific interest. For example, do networks change their structure as they operate? Do networks grow or shrink with learning? Progressive improvements in methodology, in particular the recent development of fast and automated spike-sorting methods for extracting single neuron traces from raw datasets, have greatly increased the usefulness of VSD imaging for investigating such topics in suitable invertebrate preparations.

Frequently Asked Questions

Q: Where do VSD signals come from, neurons or glia?

A: In cortex, the VSD signal is an integrated signal from all stained membranes. While both neurons and glial cells have large areas of stained membrane, their signals can be distinguished by their time course. Neuronal signals are faster (e.g., ~25-Hz oscillations); glial signals are much slower (in the 0.1- to 1-Hz range).

Q: In cortical tissue, what are the contributions of neuronal spikes versus excitatory/inhibitory postsynaptic potentials (EPSP/IPSPs)?

A: VSD signal linearly reflects the membrane potential change. In the population signal, the VSD signal amplitude is approximately the product of change in membrane potential × the time course of the change × the area of affected membrane. For each spike, its contribution to the population signal is ~100 mV × 1 ms × 1 neuron. However, the contribution of the EPSP caused by this spike can be much larger because this spike may cause EPSP on ~1,000 postsynaptic neurons. The EPSP is ~0.1 mV × 5 ms × 1,000 neurons, five times larger.

Q: Can I see an individual neuron's spikes in cortical slices?

A: You can't see individual spikes in cortical slices using the methods described in this chapter. However, if you inject dye into a single neuron, you can see individual spikes from that neuron. (See Holthoff et al. 2010.)

Q: Can I see hyperpolarizations of the neuronal population or IPSPs?

A: Yes. However, the IPSP may be difficult to distinguish from hyperpolarization of the population. Also, in cortex, inhibition is a mixture of IPSP and shunting reduction of the EPSPs.

Q: I have a fast CCD camera that can sample 2,000 frames per second. Can I use it for VSD imaging with absorbance dyes?

A: Probably not. Most CCD/CMOS cameras have 12- to 14-bit dynamic range, inadequate for detecting 0.01% changes in light intensity. In addition, most CCD/CMOS cameras have a small well size that will saturate with the high light intensities used for absorption measurements. However, a specially designed high-light Redshirt Imaging CMOS camera provides a sufficiently large well size (100 million photoelectrons) to handle the high light levels used in absorption measurements. Pixel binning of this camera allows for resolving 0.01% of fractional change with a 14-bit digital resolution.

Q: Can I use two-photon microscopy to measure VSD signals?

A: Probably not. While a high scanning rate can be achieved by line scan or local area scanning, the low light flux (10^{-3}–10^{-6} lower than single photon imaging) limits the signal-to-noise ratio. Current two-photon microscopy can only see signals larger than 1% of the fractional change, inadequate for most VSD signals.

References

Agmon, A., & Connors, B. W. (1992). Correlation between intrinsic firing patterns and thalamocortical synaptic responses of neurons in mouse barrel cortex. *J Neurosci* 12: 319–329.

Albowitz, B., & Kuhnt, U. (1993). Evoked changes of membrane potential in guinea pig sensory neocortical slices: an analysis with voltage-sensitive dyes and a fast optical recording method. *Exp Brain Res* 93: 213–225.

Bai, L., Huang, X., Yang, Q., & Wu, J. Y. (2006). Spatiotemporal patterns of an evoked network oscillation in neocortical slices: coupled local oscillators. *J Neurophysiol* 96: 2528–2538.

Bao, W., & Wu, J. Y. (2003). Propagating wave and irregular dynamics: spatiotemporal patterns of cholinergic theta oscillations in neocortex in vitro. *J Neurophysiol* 90: 333–341.

Briggman, K. L., & Kristan, W. B. Jr. (2006). Imaging dedicated and multifunctional neural circuits generating distinct behaviors. *J Neurosci* 26: 10925–10933.

Brown, C. E., Aminoltejari, K., Erb, H., et al. (2009). In vivo voltage-sensitive dye imaging in adult mice reveals that somatosensory maps lost to stroke are replaced over weeks by new structural and functional circuits with prolonged modes of activation within both the peri-infarct zone and distant sites. *J Neurosci* 29: 1719–1734.

Clarac, F., & Pearlstein, E. (2007). Invertebrate preparations and their contribution to neurobiology in the second half of the 20th century. *Brain Res Rev* 54: 113–161.

Cohen, L., Hopp, H. P., Wu, J. Y., et al. (1989). Optical measurement of action potential activity in invertebrate ganglia. *Annu Rev Physiol* 51: 527–541.

Cohen, L. B., Keynes, R. D., & Hille B. (1968). Light scattering and birefringence changes during nerve activity. *Nature* 218: 438–441.

Cohen, L. B., & Lesher, S. (1986). Optical monitoring of membrane potential: methods of multisite optical measurement. *Soc Gen Physiol Ser* 40: 71–99.

Colom, L. V., & Saggau, P. (1994). Spontaneous interictal-like activity originates in multiple areas of the CA2-CA3 region of hippocampal slices. *J Neurophysiol* 71: 1574–1585.

Demir, R., Haberly, L. B., & Jackson, M. B. (1999). Sustained and accelerating activity at two discrete sites generate epileptiform discharges in slices of piriform cortex. *J Neurosci* 19: 1294–1306.

Demir, R., Haberly, L. B., & Jackson, M. B. (2000). Characteristics of plateau activity during the latent period prior to epileptiform discharges in slices from rat piriform cortex. *J Neurophysiol* 83: 1088–1098.

Ermentrout, G. B., & Kleinfeld, D. (2001). Traveling electrical waves in cortex: insights from phase dynamics and speculation on a computational role. *Neuron* 29: 33–44.

Ferezou, I., Bolea, S., & Petersen, C. C. (2006). Visualizing the cortical representation of whisker touch: voltage-sensitive dye imaging in freely moving mice. *Neuron* 50: 617–629.

Frost, W. N., Wang, J., & Brandon, C. J. (2007). A stereo-compound hybrid microscope for combined intracellular and optical recording of invertebrate neural network activity. *J Neurosci Methods* 162: 148–154.

Frost, W. N., Wang, J., Brandon, C. J., et al. (2010). Use of fast-responding voltage-sensitive dyes for large-scale recording of neuronal spiking activity with single-cell resolution. In M. Canepari & D. Zecevic (Eds.), *Membrane potential imaging in the nervous system: Methods and applications* (pp. 53–60). New York: Springer Press.

Gao, X., Xu, W., Wang, Z., et al. (2012). Interactions between two propagating waves in rat visual cortex. *Neuroscience* 216: 57–69.

Grinvald, A., & Hildesheim, R. (2004). VSDI: a new era in functional imaging of cortical dynamics. *Nat Rev Neurosci* 5: 874–885.

Grinvald, A., Hildesheim, R., Farber, I. C., & Anglister, L. (1982). Improved fluorescent probes for the measurement of rapid changes in membrane potential. *Biophys J* 39: 301–308.

Grinvald, A., Manker, A., & Segal, M. (1982). Visualization of the spread of electrical activity in rat hippocampal slices by voltage-sensitive optical probes. *J Physiol* 333: 269–291.

Gupta, R. K., Salzberg, B. M., Grinvald, A., et al. (1981). Improvements in optical methods for measuring rapid changes in membrane potential. *J Membr Biol* 58: 123–137.

Hill, E. S., Bruno, A. M., & Frost, W. N. (2014) Recent developments in VSD imaging of small neuronal networks. *Learn Mem* 21: 499–505.

Hill, E. S., Moore-Kochlacs, C., Vasireddi, S. K., et al. (2010). Validation of independent component analysis for rapid spike sorting of optical recording data. *J Neurophysiol* 104: 3721–3731.

Hill, E. S., Vasireddi, S. K., Bruno, A. M., et al. (2012). Variable neuronal participation in stereotypic motor programs. *PLoS ONE* 7: e40579.

Hirata, Y., & Sawaguchi, T. (2008). Functional columns in the primate prefrontal cortex revealed by optical imaging in vitro. *Neurosci Res* 61: 1–10.

Hirota, A., Sato, K., Momose-Sato, Y., et al. (1995). A new simultaneous 1020-site optical recording system for monitoring neural activity using voltage-sensitive dyes. *J Neurosci Methods* 56: 187–194.

Holthoff, K., Zecevic, D., & Konnerth, A. (2010). Rapid time course of action potentials in spines and remote dendrites of mouse visual cortex neurons. *J Physiol* 588 (Pt 7): 1085–1096.

Huang, X., Troy, W. C., Yang, Q., et al. (2004). Spiral waves in disinhibited mammalian neocortex. *J Neurosci* 24: 9897–9902.

Huang, X., Xu, W., Liang, J., et al. (2010). Spiral wave dynamics in neocortex. *Neuron* 68: 978–990.

Hume, R. I., Getting, P. A., & Del Beccaro, M. A. (1982). Motor organization of *Tritonia* swimming. I. Quantitative analysis of swim behavior and flexion neuron firing patterns. *J Neurophys* 47: 60–74.

Jalife, J. (2003) Rotors and spiral waves in atrial fibrillation. *J Cardiovasc Electrophysiol* 14: 776–780.

Jin, W., Zhang, R. J., & Wu, J. Y. (2002). Voltage-sensitive dye imaging of population neuronal activity in cortical tissue. *J Neurosci Meth* 115: 13–27.

Kleinfeld, D., & Delaney, K. R. (1996). Distributed representation of vibrissa movement in the upper layers of somatosensory cortex revealed with voltage-sensitive dyes. *J Comp Neurol* 375: 89–108.

Laaris, N., Carlson, G. C., & Keller, A. (2000). Thalamic-evoked synaptic interactions in barrel cortex revealed by optical imaging. *J Neurosci* 20: 1529–1537.

Lippert, M. T., Takagaki, K., Xu, W., et al. (2007). Methods for voltage-sensitive dye imaging of rat cortical activity with high signal-to-noise ratio. *J Neurophysiol* 98: 502–512.

Loew, L. M., Cohen, L. B., Dix, J., et al. (1992). Naphthyl analog of the aminostyryl pyridinium class of potentiometric membrane dyes shows consistent sensitivity in a variety of tissue, cell, and model membrane preparations. *J Membr Biol* 130: 1–10.

London, J. A., Cohen, L. B., & Wu, J. Y. (1989). Optical recordings of the cortical response to whisker stimulation before and after the addition of an epileptogenic agent. *J Neurosci* 9: 2182–2190.

London, J. A., Zecevic, D., & Cohen, L. B. (1987). Simultaneous optical recording of activity from many neurons during feeding in Navanax. *J Neurosci* 7: 649–661.

Longley, R. D. (1984). Axon surface infolding and axon size can be quantitatively related in gastropod molluscs. *J Exp Biol* 108: 163–177.

Lukatch, H. S., & MacIver, M. B. (1997). Physiology, pharmacology, and topography of cholinergic neocortical oscillations in vitro. *J Neurophysiol* 77: 2427–2445.

Ma, H. T., Wu, C. H., & Wu, J. Y. (2004). Initiation of spontaneous epileptiform events in the rat neocortex in vivo. *J Neurophysiol* 91: 934–945.

MacLean, J. N., Fenstermaker, V., Watson, B. O., & Yuste, R. (2006). A visual thalamocortical slice. *Nat Methods* 3: 129–134.

Miller, E. W., Lin, J. Y., Frady, E. P., et al. (2012). Optically monitoring voltage in neurons by photo-induced electron transfer through molecular wires. *Proc Natl Acad Sci USA* 109: 2114–2119.

Mirolli, M., & Talbott, S. R. (1972). The geometrical factors determining the electrotonic properties of a molluscan neurone. *J Physiol* 227: 19–34.

Mochida, H., Sato, K., Arai, Y., et al. (2001). Optical imaging of spreading depolarization waves triggered by spinal nerve stimulation in the chick embryo: possible mechanisms for large-scale coactivation of the central nervous system. *Eur J Neurosci* 14: 809–820.

Orbach, H. S., Cohen, L. B., & Grinvald, A. (1985). Optical mapping of electrical activity in rat somatosensory and visual cortex. *J Neurosci* 5: 1886–1895.

Petersen, C. C. (2007). The functional organization of the barrel cortex. *Neuron* 56: 339–355.

Petersen, C. C., Grinvald, A., & Sakmann, B. (2003). Spatiotemporal dynamics of sensory responses in layer 2/3 of rat barrel cortex measured in vivo by voltage-sensitive dye imaging combined with whole-cell voltage recordings and neuron reconstructions. *J Neurosci* 23: 1298–1309.

Petersen, C. C., Hahn, T. T., Mehta, M., et al. (2003). Interaction of sensory responses with spontaneous depolarization in layer 2/3 barrel cortex. *Proc Natl Acad Sci USA* 100: 13638–13643.

Popovic, M. A., Foust, A. J., McCormick, D. A., & Zecevic, D. (2011). The spatio-temporal characteristics of action potential initiation in layer 5 pyramidal neurons: a voltage imaging study. *J Physiol* 589: 4167–4187.

Prechtl, J. C., Cohen, L. B., Pesaran, B., et al. (1997). Visual stimuli induce waves of electrical activity in turtle cortex. *Proc Natl Acad Sci USA* 94: 7621–7626.

Ross, W. N., Salzberg, B. M., Cohen, L. B., et al. (1977). Changes in absorption, fluorescence, dichroism, and birefringence in stained giant axons: optical measurement of membrane potential. *J Membr Biol* 33: 141–183.

Sattelle, D. B., & Buckingham, S. D. (2006). Invertebrate studies and their ongoing contributions to neuroscience. *Invert Neurosci* 6: 1–3.

Senseman, D. M., Vasquez, S., & Nash, P. L. (1990). Animated pseudocolor activity maps PAMs: Scientific visualization of brain electrical activity. In D. Schild (Ed.), *Chemosensory information processing* (pp. 329–347). NATO ASI Series, Vol. H39. Berlin: Springer-Verlag.

Shoham, D., Glaser, D. E., Arieli, A., et al. (1999). Imaging cortical dynamics at high spatial and temporal resolution with novel blue voltage-sensitive dyes. *Neuron* 24: 791–802.

Slovin, H., Arieli, A., Hildesheim, R., & Grinvald, A. (2002). Long-term voltage-sensitive dye imaging reveals cortical dynamics in behaving monkeys. *J Neurophysiol* 88: 3421–3438.

Tanifuji, M., Sugiyama, T., & Murase, K. (1994). Horizontal propagation of excitation in rat visual cortical slices revealed by optical imaging. *Science* 266: 1057–1059.

Tasaki, I., Watanabe, A., Sandlin, R., & Carnay, L. (1968). Changes in fluorescence, turbidity, and birefringence associated with nerve excitation. *Proc Natl Acad Sci USA* 61: 883–888, 1968.

Tsau, Y., Guan, L., & Wu, J. Y. (1998). Initiation of spontaneous epileptiform activity in the neocortical slice. *J Neurophysiol* 80: 978–982.

Tsau, Y., Guan, L., & Wu, J. Y. (1999). Epileptiform activity can be initiated in various neocortical layers: an optical imaging study. *J Neurophysiol* 82: 1965–1973.

Winfree, A. T. (2001). *The geometry of biological time.* New York: Springer.

Wu, J. Y., Falk, C. X., Cohen, L., et al. (1993). Optical measurement of action potential activity in invertebrate ganglia. *Jpn J Physiol* 43(Suppl 1): S21–29.

Wu, J. Y., Guan, L., Bai, L., & Yang, Q. (2001). Spatiotemporal properties of an evoked population activity in rat sensory cortical slices. *J Neurophysiol* 86: 2461–2474.

Wu, J. Y., Guan, L., & Tsau, Y. (1999). Propagating activation during oscillations and evoked responses in neocortical slices. *J Neurosci* 19: 5005–5015.

Wu, J. Y., London, J. A., Zecevic, D., et al. (1988). Optical monitoring of activity of many neurons in invertebrate ganglia during behaviors. *Experientia* 44: 369–376.

Xu, W., Huang, X., Takagaki, K., & Wu, J. Y. (2007). Compression and reflection of visually evoked cortical waves. *Neuron* 55: 119–129.

Zecevic, D., Wu, J., Cohen, L.B., et al. (1989). Hundreds of neurons in the *Aplysia* abdominal ganglion are active during the gill-withdrawal reflex. *J Neurosci* 9: 3681–3689.

Optogenetics and Electrophysiology

Hua-an Tseng, Richie E. Kohman, and Xue Han

Introduction

Optogenetics combines optical and genetic techniques to manipulate neural activity. Genetic techniques are used to transduce neurons of interest in the brain with light-sensitive proteins, rendering them sensitive to light. Specific wavelengths of light can then be pulsed on the transduced neurons to activate or silence them on the millisecond time scale.

Since the first demonstration of using pulses of blue light to drive activity in neurons expressing the light-activated microbial rhodopsin, channelrhodopsin 2 (Boyden et al. 2005), optogenetics has developed rapidly and its use has become widespread. Optogenetics is based on the discovery of native light sensor proteins in archaebacteria and green algae. These small microbial opsins, whose gene size is similar to that of green fluorescent protein (GFP, 1 kilobase pair), can easily be expressed in cells of interest using genetic techniques and activated or deactivated by application of light, thus regulating movement of ions across the cell membrane. So far, three major classes of microbial opsins (channelrhodopsins, halorhodopsins, and archaerhodopsins) have been developed (Fig. 9.1) (Han 2012a). These have been used in specific cells in many model systems such as *Caenorhabditis elegans*, rodents, and monkeys. The translational potential of optogenetics has also been highlighted by studies in human retina explant (Busskamp et al. 2012). Optogenetics has also made it possible to explore of the role of specific populations of neurons in computation and disease. A number of recent reviews and books have summarized various aspects of the current state of the field (i.e., Bernstein and Boyden 2011, Chow et al. 2012, Han 2012b, Knopfel and Boyden 2012, Miesenbock 2011, Yizhar et al. 2011a, Zhang et al. 2011). Here, we give a brief overview of the technology, highlighting major technical considerations and providing some examples in which optogenetics is

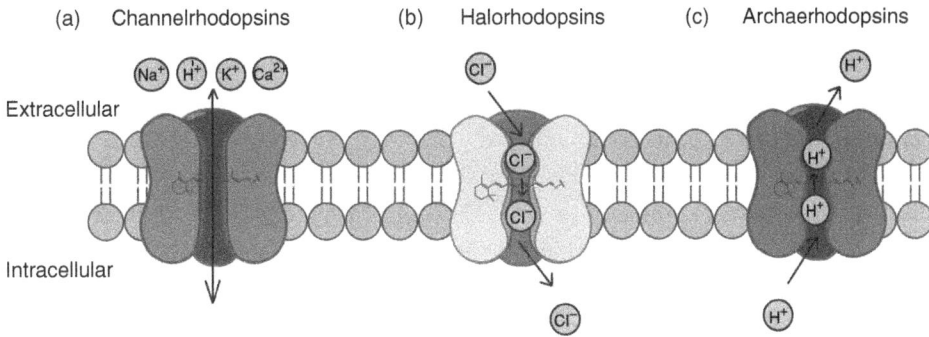

FIGURE 9.1 Optogenetic molecular sensors. Upon light illumination, channelrhodopsins passively transport Na$^+$, K$^+$, H$^+$, and Ca^{2+} down their electrochemical gradients to depolarize neurons (A); halorhodopsins actively pump Cl$^-$ into the cell to hyperpolarize neurons (B); and archaerhodopsins actively pump H$^+$ out of the cell to hyperpolarize neurons (C).

used to map neural circuits *in vivo*. Specifically, we will address (1) considerations in choosing the proper optogenetic molecular sensors, (2) genetic transduction methods to express opsins in cells of interest, (3) the assembly of appropriate optical and electrical hardware for illuminating neurons, (4) examples applying these technologies in various model systems, and (5) combining optogenetics with electrophysiology techniques.

Optogenetic Molecular Sensors

The key components of optogenetics are small, single-compartment light-activated proteins called microbial opsins. An opsin protein along with its covalently bound cofactor, retinal, is called a rhodopsin. Type I microbial opsins are a group of photoactive seven-transmembrane proteins found in archaea, bacteria, algae, and fungi. Since the 1970s, microbial opsins have been widely studied because of their critical roles in light-sensing and photosynthetic functions; however, it was not until recently that scientists successfully used them to control neural activity.

In general, there are two groups of microbial opsins: transport rhodopsins and sensory rhodopsins (Spudich et al. 2000). During light stimulation, transport rhodopsins actively drive ions across plasma membranes. These ions include chloride (halorhodopsin) or protons (bacteriorhodopsins, archaerhodopsins, proteorhodopsins). Sensory rhodopsins, on the other hand, do not have ion transport capabilities; instead, most of them activate intracellular signaling pathways. However, one sensory rhodopsin, channelrhodopsin, functions as a light-gated ion channel. At high light intensity, channelrhodopsin acts as a cation channel, while at low light intensity, it acts as a calcium channel (Sineshchekov et al. 2009).

Each type of rhodopsin has its own efficacy in controlling the membrane potential of cells, mostly depending on the cycle of photochemical events (photocycle) induced by each photon. However, all opsins used for optogenetics share certain mechanistic features. During light stimulation, rhodopsins undergo a series of intermediate states, and the kinetics of the transitions between these states determines the response time of the opsin. In the dark state, the chromophore, all-trans retinal (vitamin A-aldehyde), forms a protonated Schiff base with the conserved lysine residue in the seventh trans-membrane domain of the opsin (Spudich et al. 2000). Upon absorption of a photon, the all-trans retinal transforms into the 13-cis form (a process known as photoisomeriza-tion) and induces sequential conformational changes in the opsin. The conformational changes involve multiple intermediate states and result in the transport of ions across the cell membrane. At the end of the cycle, 13-cis retinal isomerizes back to all-trans retinal as the opsin finishes its photocycle.

Channelrhodopsins: Optogenetic Activators

Channelrhodopsins are a group of light-gated ion channels found naturally in green algae. They share homology with phototaxis receptors and light-driven ion transport-ers in many prokaryotes. In 2005, Boyden et al. first demonstrated that neurons were able to respond to blue light pulses after expressing channelrhodopsin-2 from the green alga *Chlamydomonas reinhardtii* (ChR2, the N-terminal domain of the native wide-type form, 315 amino acids) (Boyden et al. 2005). Although ChR2 behaves as an ion channel during blue light pulses, a recent study showed that it is actually a leaky proton pump (Feldbauer et al. 2009). Upon light illumination, ChR2 opens like a channel, allowing mono- and divalent cations, Na^+, K^+, H^+, and Ca^{2+} ions, to flow through the pore (Nagel et al. 2003).

As a gated ion channel, ChR2 requires energy only when opening the channel. The photonic energy provided by blue light initiates conformational changes required to open the channel. After that, ions passively diffuse through, following a concentration gradient without any further energy consumption. Therefore, ChR2 has better energy efficiency compared to light-gated ion pumps, such as halorhodopsin and archaerho-dopsin. However, ChR2 has a smaller single-channel conductance than most of the endogenous ion channels (Feldbauer et al. 2009, Lin et al. 2009, Nagel et al. 2003), and thus a relatively high ChR2 expression is usually necessary to activate a neuron.

The success of using ChR2 as an optogenetic sensor prompted much effort in improving its functions, such as increasing photocurrent amplitude, altering channel opening and closing kinetics, and shifting the activation color spectrum. For example, ChR2 point mutants ChR2 (H134R), ChR2 (T159C), and several chimeras of ChR1 and ChR2 (ChIEF, ChRGR) exhibit higher-amplitude photocurrent (Berndt et al. 2011, Lin et al. 2009, Nagel et al. 2005, Wang et al. 2009, Wen et al. 2010). CheTA shows faster kinetics synergistic with higher-amplitude photocurrent (Gunaydin et al. 2010). Point mutants ChR2 (D156A) and ChR2 (C128A/S) show slower kinetics and thus allow sus-tained long-term depolarization lasting minutes (Bamann et al. 2010, Berndt et al. 2009). VChR1 from *Volvox carteri*, MChR1 from *Mesostigma viride*, and the chimeric C1V1 possess red-shifted absorbance spectra (Govorunova et al. 2011, Yizhar et al. 2011b, Zhang et al. 2008), whereas ChR2 (L132C) exhibits enhanced calcium permeability

(Kleinlogel et al. 2011). As the toolbox of ChR2 variants becomes more complete, scientists can choose a modified form based on their experimental needs to examine the roles of specific neurons in neural circuit function and pathologies.

Halorhodopsins: Optogenetic Inhibitors

In 2007 the microbial opsin halorhodopsin from *Natronomonas pharaonis* (Halo/NpHR) was employed to rapidly and reversibly silence neural activity (Han and Boyden 2007, Zhang et al. 2007). Halorhodopsins are light-activated chloride pumps naturally found in archaebacteria. Although the illumination of halorhodopsins can silence cultured neurons in well-controlled patterns, the photocurrents are not sufficient to mediate strong inhibition *in vivo*. This is largely due to poor membrane localization when halorhodopsin is expressed in mammalian neurons. Improvements in targeting Halo to the plasma membrane have been made by attaching membrane-targeting sequences. The recombinant proteins (Halo2.0/eNpHR and Halo3.0/eNpHR3.0) can mediate much larger photocurrents to provide stronger silencing *in vivo* (Gradinaru et al. 2008, 2010, Ma et al. 2001, Zhao et al. 2008).

Besides poor membrane localization, another disadvantage of halorhodopsin is its long recovery time after extended yellow light stimulation, mainly because of its nonconducting intermediate states (Bamberg et al. 1993, Han and Boyden 2007). Blue light illumination was found to shorten the recovery time from several minutes to merely a few milliseconds (Chow et al. 2010, Han and Boyden 2007). Finally, the constant transport of chloride ions via halorhodopsin could potentially alter extracellular and/or intracellular chloride concentrations, which could shift the chloride electrochemical gradient, critical for the reversal potential of the GABA receptor. For example, prolonged activation of hippocampal neurons expressing Halo3.0/eNpHR3.0 has been shown to alter inhibitory GABA synaptic current, resulting in an increased neural excitability that lasts well after Halo3.0 action (Raimondo et al. 2012).

Archaerhodopsins: Optogenetic Inhibitors

Archaerhodopsin-3 (Arch) from *Halorubrum sodomense* was discovered in 2010 to rapidly and reversibly silence neurons by pumping protons out of a cell (Chow et al. 2010). Archaerhodopsins are light-activated proton pumps, which express well on the plasma membrane and can generate strong outward inhibitory photocurrents. Several variants of archaerhodopsins have since been developed, including ArchT from *Halorubrum* strain TP009 with an improved photocurrent that can silence larger brain volumes (Han et al. 2011) and Mac from *Leptosphaeria maculans* with a blue-shifted action spectrum (Chow et al. 2010). Continued pumping of protons by Arch has been shown not to alter cellular pH beyond physiological limits, likely due to intrinsic compensatory mechanisms. However, it remains to be established whether constant outward proton flow acidifies the local extracellular environment, which could influence the excitability of surrounding neurons not expressing Arch and leading to secondary neural modulation effects.

Other Optogenetic Sensors

Light-activated animal (type II) opsins have been used to sensitize neurons to light, such as the three-protein drosophila photoreceptor system (Zemelman et al. 2002), rat rhodopsin 4 (RO4) (Li et al. 2005), and melanopsin (Ye et al. 2011). In addition, by replacing the intracellular domains of microbial opsins with the intracellular domains of G-protein–coupled receptors, G-protein–coupled receptors have been sensitized to light (Airan et al. 2009). The ability to control different patterns of ionic flux and signaling pathways expands the diversity of optogenetic control of different cellular functions.

A number of other strategies involving the combination of chemical and genetic methods have also been developed to sensitize cells to light, including the optical uncaging of chemical ligands of heterologously expressed receptors (Lima and Miesenbock 2005, Zemelman et al. 2003) and the photo-switching of chemical bonds between modified chemical ligands and mutated ion channels (Banghart et al. 2004, Janovjak et al. 2010). However, the use of these approaches *in vivo* is limited by the need to introduce exogenous chemicals.

Factors that Affect Optogenetic Control of Neurons

Each type of opsin has distinctive molecular properties, and these properties play important roles in determining their efficacy in controlling neurons. For example, the kinetics of opsin switching between intermediate states determines the kinetics of the photocurrent, while the expression level of an opsin determines the amplitude of total current generated by light stimulation.

Three major properties of opsins influence optogenetic control: total photocurrent amplitude, response time, and action spectrum.

Photocurrent amplitude. The total photocurrent is the combined result of the ion transport speed of an individual opsin molecule, the overall expression level of opsin on a membrane, and the intensity and efficiency of light irradiation. In general, channelrhodopsins can generate larger photocurrents than halorhodopsins and archaerhodopsins. This is because channelrhodopsins use photonic energy to open the channels, whereas halorhodopsins and archaerhodopsins constantly consume energy to transport ions across the cell membrane. Nonetheless, all three classes of microbial opsins have been engineered to provide sufficient control of neuronal excitability *in vivo* to influence behavior across a number of species.

Response time. The response time to light stimulation of an opsin is mainly determined by the kinetics of conformational changes between different intermediate states. Because the transitions between states require photonic energy, halorhodopsins and archaerhodopsins have better response times to light stimulation and can switch between on and off states quickly. The intermediate states of the channelrhodopsin photocycle, on the other hand, do not rely on photonic energy except for the initiation step. Therefore, the channel closing rate of channelrhodopsin variants can vary from a few milliseconds to several minutes (Yizhar et al. 2011a). Prolonged activation of opsins, however, may lead to an overall increase in membrane conductance, making it harder to depolarize or hyperpolarize a neuron.

Action spectrum. Most opsins have broad action spectrums. For example, ChR2 has an absorbance maximum at 470 nm (blue light), but any light between 430 and 520 nm can activate ChR2 with more than 50% efficiency (Nagel et al. 2003). Arch has a peak response to green light but also activates with more than 50% efficiency with light of 500 to 600 nm (Chow et al. 2010, Duschl et al. 1990, Han and Boyden 2007). Therefore, to develop a multicolor optogenetic system, it is important to create opsins with not only distinct peak activation wavelengths but also narrower action spectrums. Among all possibilities, an opsin responding to infrared light could have the additional advantage of being able to be activated in a deep brain area without the implantation of an optical fiber because infrared light could penetrate brain tissue better.

Genetic Transduction of Cells with Opsins

A major advantage of optogenetics is the ability to selectively control the activity of specific cell types. This is achieved by taking advantage of genetic regulatory elements that are unique to certain neuronal populations. Major gene transfer techniques include viral-based gene delivery and whole animal transgenic approaches. Nonviral-based gene delivery methods exhibit low transduction efficiency *in vivo* and will not be further discussed here (Luo and Saltzman 2000).

Viral Gene Delivery

In genetically intractable species, viral gene delivery remains the most widely used method to transduce brain cells. With little toxicity and excellent transduction efficiency, lentivirus and adeno-associated virus (AAV) are the most commonly used viral vectors for basic research and gene therapy trials (Han 2012b, Waehler et al. 2007). AAV and lentivirus have been used to express rhodopsins in a variety of model systems, including mice, rats, monkeys, and human retina explant (Busskamp et al. 2012, Zhang et al. 2010).

The disadvantage with using viral approaches is that viruses have limited DNA packaging ability, with AAV vectors capable of packaging ~4.7 kb of DNA and lentiviral vectors ~8 kb. This limitation has made it difficult to restrict opsin expression to target cells using a promoter-based strategy, as some promoters can be much larger than 4–8 kb. In addition, the intrinsic tropism of viruses may further influence gene transduction efficiency across different cell types in different brain regions. In general, AAV2, 5, 8, and 9 and lentivirus have been shown to effectively infect cells in a variety of brain regions (Han 2012b).

Transgenic Approach

Using the transgenic approach, opsin genes can be directly engineered into the chromosome to achieve targeted expression in specific cell types. However, such approaches require making an individual transgenic animal for each cell type and each opsin of interest. Alternatively, one can take advantage of phage-derived Cre-LoxP recombination technology (Sauer and Henderson 1988, Tsien et al. 1996). Cre recombinase recognizes two 34-bp DNA sequences called LoxP sites and removes the DNA fragment between them via recombination (Fig. 9.2A). By selectively expressing Cre in specific cell types, one can restrict opsin expression into these Cre-expressing cells by injecting a virus containing LoxP-flanked opsin genes or by crossing the Cre-expressing mouse with a transgenic mouse that expresses LoxP-flanked opsin genes. This is a powerful technique because there are a large number of transgenic mice with targeted Cre expression in many cell types.

Anatomical Pathway-Specific Expression of Opsins

One may desire to investigate the function of certain anatomical pathways with optogenetics. If the cell bodies and the synaptic terminals for the pathway are spatially well

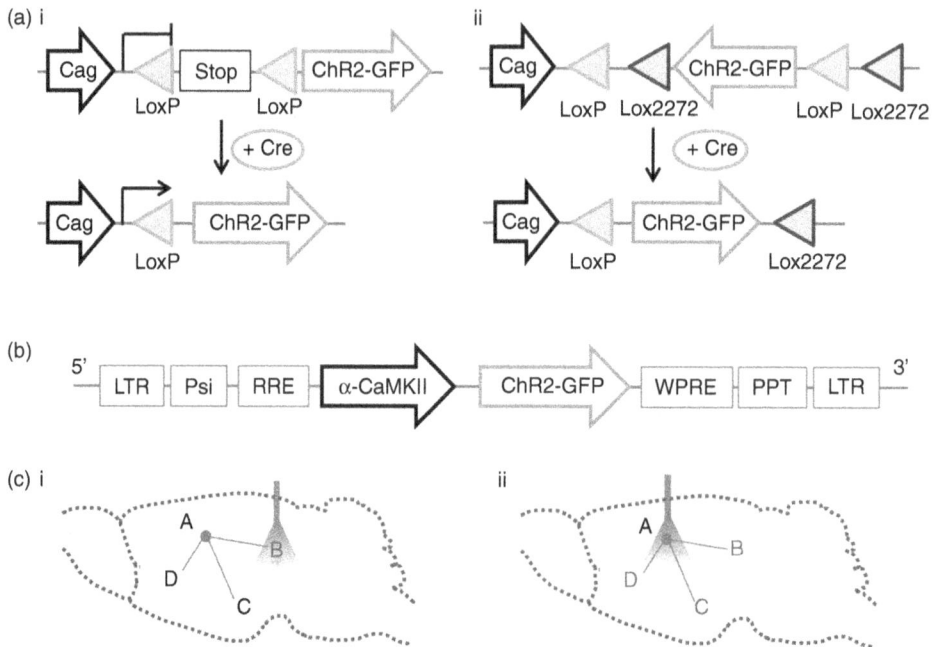

FIGURE 9.2 **Cell-specific targeting.** A: Cell type-specific targeting with Cre-LoxP strategies. Ai: An example demonstrating targeted expression of opsins in Cre-expressing cells by removing transcription/translation stop sequences that block the expression of opsins. Aii: An example demonstrating targeted expression of opsins in Cre-expressing cells by flipping the opsin gene that is otherwise in the noncoding orientation (adapted from Atasoy et al. 2008). B: Cell type-specific targeting using short promoters in a virus. An example demonstrating selective targeting of cortical excitatory neurons using a lentivirus with a CamKII promoter (adapted from Han et al. 2009). C: Anatomical pathway-specific targeting. Ci: An example demonstrating light illumination at area B selectively activating an A-to-B projection. Cii: An example demonstrating light illumination of neuron cell bodies that were retrogradely labeled through axon terminals projecting to area B. In this case, all the downstream areas that receive projections from area A are modulated.

separated, one can express opsins in the cell bodies and then illuminate the terminals for optogenetic controls (see Fig. 9.2Ci). This strategy has been demonstrated with ChR2 *in vitro* (Cruikshank et al. 2010, Petreanu et al. 2007) and *in vivo* (Tye et al. 2011) but requires sufficient expression of opsins at the axon terminals. One potential concern is that activation of the axon terminals, and fibers of passage in the illuminated area, may lead to antidromic stimulation, a similar concern as presented with electrical stimulation.

To selectively label cells projecting to a brain region of interest, one could use a virus capable of retrograde labeling. One lentivirus, EIAV pseudotyped with rabies glycoprotein, has shown promise to retrogradely label cells (Mazarakis et al. 2001), but it remains to be determined whether such a strategy can express sufficient opsin proteins for effective optogenetic modulation. Engineered rabies virus can effectively label in a retrograde fashion but may ultimately lead to neuronal dysfunction, probably due to extensive gene expression (Wickersham et al. 2010). Another disadvantage of this strategy is that if the optogenetically targeted neurons project to multiple downstream areas, all of these areas will be perturbed during light stimulation, not just the areas of interest (see Fig. 9.2Cii).

Light Illumination of Transduced Neurons

Once neurons are genetically transduced to express desired opsins, proper light illumination at a specific wavelength is needed to control the transduced neurons optically. As light stimulation and measuring methods could create a variety of types of cell damage, it is important to design correct control experiments to rule out possible artifacts. For example, when a virus is used to transduce neurons with fluorescently tagged opsin proteins, a proper control would be to use the same type of virus to transduce a similar set of neurons with only the fluorescent protein.

Light Wavelength Considerations

Proper wavelength of light illumination is required for a given opsin. In the brain, blood hemoglobins are the major factors for light absorption and scattering (Mobley and Vo-Dinh 2003) (Fig. 9.3). Monte Carlo simulation provides a good estimate of light propagation patterns within brain tissue, and it was estimated that at the visible light wavelengths where most opsins operate, light falls off nonlinearly with the sharpest intensity drop within the first several hundred micrometers (Adamantidis et al. 2007, Bernstein et al. 2008, Chow et al. 2010, Han et al. 2009). It is thus advantageous to develop opsins that are operated by red light (i.e., >630 nm), where absorption by hemoglobin and oxygenated hemoglobin absorption is drastically reduced. In addition, opsins possessing narrower absorption spectra are desirable for multicolor optogenetic experiments.

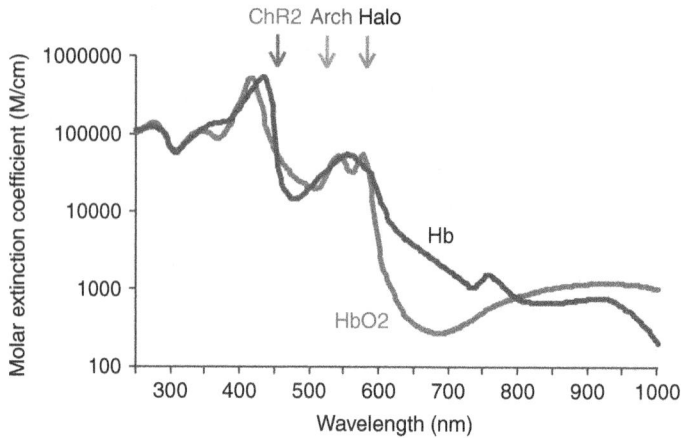

FIGURE 9.3 Hemoglobin, oxygenated (HbO2) and deoxygenated (Hb) absorption spectrum. Values are based on that summarized by Scott Prahl, http://omlc.ogi.edu/spectra. The peak excitation wavelengths for ChR2, Arch, and Halo are indicated.

Tissue Damage from Device Insertion and Heat

Device insertion as well as heat generated by light can damage tissue. To reduce mechanical damage from device insertion, ideally one would use the smallest devices possible. For example, multiple thin fibers or fiber arrays are helpful in reducing mechanical tissue damage (Bernstein et al. 2012). Heat-generated damage is hard to evaluate. Heating alone has been shown to influence neural activity, as demonstrated by pulsed infrared light stimulation (Wells et al. 2005) or possibly by pulsed ultrasound stimulation (Bernstein et al. 2012, Tufail et al. 2011). Light power in the range of a few milliwatts seems to be safe *in vivo*. Again, proper controls (e.g., repeating the experiment with light stimulation in the absence of opsin expression) are necessary to ensure that photostimulation effects are not due to heat.

Optical Artifacts on Metal Electrodes

A major advantage of optogenetics is that it is compatible with many readout technologies, such as electrophysiology, functional magnetic resonance imaging (fMRI), behavioral testing, and cellular imaging. However, laser light produces a strong voltage deflection artifact when directed to the tip of a metal electrode (Fig. 9.4) (Ayling et al. 2009, Han et al. 2009, 2011). This type of artifact is not seen with glass microelectrodes where the silver/silver chloride electrode in the glass is not exposed to light. The light-induced artifact in metal electrodes is slow to evolve and thus corrupts local field potential (LFP) but not spike waveforms. Thus, caution must be exercised when interpreting LFP results in optogenetic experiments. Optimizing electrode tips, such as coating with conducting polymer indium tin oxide (ITO), may prove useful in reducing this artifact on LFP recordings (Zorzos et al. 2009).

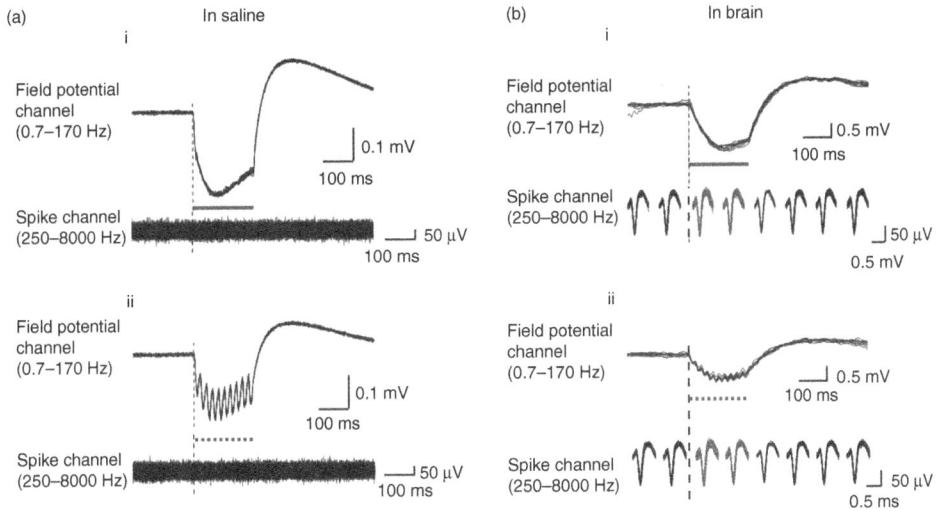

FIGURE 9.4 Light-mediated artifact on metal electrodes. Optical artifact observed on tungsten electrodes immersed in saline (A) or brain (B) upon exposure to 200-ms blue light pulses (i) or trains of 10-ms blue light pulses delivered at 50 Hz (ii). Light pulses are indicated by blue dashes. Electrode data were hardware filtered using two data acquisition channels operating in parallel, yielding a low-frequency component ("field potential channel") and a high-frequency component ("spike channel"). For the "spike channel" traces taken in brain (B), spikes were grouped into 100-ms bins, and then the binned spikes were displayed beneath the corresponding parts of the simultaneously acquired "field potential channel" signal. (Shown are the spikes in eight such bins: two bins before light onset, two bins during the light delivery period, and four bins after light cessation) (adapted from Han et al. 2009).

Combination of Optogenetics and Electrophysiology Recording

Traditionally, researchers examine the functions of neural circuits by recording neural activity with various types of electrodes (see Chapters 1–3), including a tetrode or multi-electrode device (see Chapter 4), when animals are performing behavioral tasks. With the help of optogenetics, researchers can alter neural activity, and many labs are developing devices to record neural activity while manipulating excitability through the use of optogenetics (see Yizhar et al. 2011a for a detailed description). In 2011, an "optetrode," a device combining a single optical fiber surrounded by four tetrode bundles, was used to record neural responses to the optical stimulation in the prefrontal cortex of freely moving mice (Anikeeva et al. 2012). More recently, the flexDrive, a new device with 65 channels and multiple optical fibers, demonstrated the possibility of recording and optically stimulating multiple sites at the same time (Voigts et al. 2013). These newly developed tools provide more options for researchers to investigate the neural circuits related to animal behavior.

Applications of Optogenetics in Various Model Organisms

Dynamic interactions between different types of neurons or glia, on various time scales,, ultimately influence behavior and cognition. Traditional methods such as lesions, microstimulation, or pharmacological perturbations lack the temporal precision to rapidly and reversibly manipulate specific cell types. Optogenetics enables precise perturbation of a specific set of neurons that can be genetically transduced via viruses or transgenic strategies. By pulsing certain patterns of light to activate or silence transduced cells, it is now possible to examine the roles these cells play in neural computation and pathology in all commonly used experimental animal models.

Caenorhabditis elegans

C. elegans is a simple animal model used for studying neural development and neural circuit computation. Its neural network consists of 302 neurons and is well understood. Because of the well-developed genetic toolkits that have been developed and the fact that the body is optically transparent, optogenetics can easily be performed on this organism. The use of optogenetics in *C. elegans* requires the addition of all-trans retinal (e.g., adding it to the culture medium) as the *C. elegans* nervous system does not produce a sufficient amount for the opsins to operate. When Arch is expressed in all neurons or muscles, illuminating the entire animal with green light causes it to freeze (Okazaki et al. 2012). More sophisticated experiments that illuminate a specific part of the moving animal can be achieved by using a CCD camera tracking system to control the direction of the light beam from a LED array (Kawazoe et al. 2012) or a projector (Stirman et al. 2011). For example, optical stimulation or inhibition of different sensory neurons or command interneurons in the touch circuit has revealed their respective roles in controlling movements (Husson et al. 2012, Stirman et al. 2011). In another study, optogenetic stimulation of mechanosensory neurons has demonstrated age effects on habituation (Timbers et al. 2012).

Drosophila

During its larval stage, Drosophila has a transparent body and can be easily investigated optogenetically (Honjo et al. 2012). Similar to *C. elegans*, neurons in fruit flies do not have enough endogenous all-trans retinal and need to obtain retinal from their food for opsins to function properly. ChR2 has been expressed in olfactory receptor neurons in Drosophila to study olfaction (Bellmann et al. 2010). In another study, optogenetic silencing of central pattern generator neurons using Halo/NpHR showed that activation of motor neurons in different body segments requires signals from anterior segments, and the state of locomotion can be memorized for a few seconds (Inada et al. 2011).

Zebrafish

Zebrafish also have a transparent body, allowing easy optical manipulations. A digital micromirror device has been applied to illuminate two locations at the same time (Zhu et al. 2012). In transgenic zebrafish, release of dopamine and GABA depends on the length of optogenetic stimulation of ChR2-expressing interneurons (Bundschuh et al. 2012). In another study, optogenetics has been used to locate the pacemaker cells in the cardiomyocytes to alter heart rate with various frequencies of light pulses (Arrenberg et al. 2010).

Rodent

Rodents are one of most commonly used vertebrate models in neuroscience. Unlike *C. elegans*, Drosophila, and zebrafish, rodents do not have transparent bodies; their brains are protected under the skull and therefore light delivery for optogenetics in rodents (and also in primates) often requires direct positioning of the light source on or inside the brain. For head-fixed recording, optical fibers are acutely or chronically inserted into the brain to reach target areas to deliver light. For freely moving animals in behavioral experiments, light can be delivered through a thinned skull to target neurons located at the brain surface (Drew et al. 2010) or through chronically implanted optical fibers targeting deeper brain areas (Bernstein and Boyden 2011). Optic fibers can be designed so that multiple fibers form a probe with openings at different depths along the probe axis (Zorzos et al. 2010) or an array to reach different sites (Bernstein and Boyden 2011). Furthermore, the tip of each fiber can be modified to release light in different beam shapes (Kravitz and Kreitzer 2011). The optical fiber then needs to be connected to a laser or LED through an extension optical fiber, or a directly implanted LED can be connected to a wire to provide power. In some cases, the extension optical fiber or the power cord could hinder the animal's movement and interfere with performance on behavioral tasks. A commutator is often used to reduce tension on the wires. Recently, a fully wireless LED implant has been developed to solve this problem (Wentz et al. 2011). Finally, the optical fiber can also be coupled to electrodes, known as an "optetrode" (Anikeeva et al. 2012, Stark et al. 2012, Wang et al. 2011, 2012), to record neural activity in behaving rodents during optogenetic perturbation.

Optogenetics has been successfully implemented in a variety of neural circuits to investigate the functional role of specific cells. For example, parvalbumin-positive (PV) interneurons have been inhibited or activated optogenetically, resulting in a suppression or enhancement of gamma-frequency oscillations (Cardin et al. 2009, Sohal et al. 2009). This suggests that PV neurons are important for the generation of gamma oscillations (30–70 Hz), which have been associated with attention and other cognitive functions. In another example, optogenetic stimulation of hypocretin neurons in the hypothalamus can induce sleep–wake status transitions (Adamantidis et al. 2010), and distributing sleep with optical stimulation can cause inhibition of memory consolidation (Rolls et al. 2011).

In fear learning and memory tasks, optogenetic stimulation of ChR2-expressing PV neurons in auditory cortex during a foot shock blocks fear learning, and disinhibition of the pyramidal neurons in auditory cortex is required for associations between foot shock and sound (Letzkus et al. 2011). In another task where rodents were trained to self-administrator drugs, a model system for studying drug addiction, optogenetic silencing of cholinergic neurons in nucleus accumbens reduced the effect of cocaine conditioning (Witten et al. 2010). Optogenetic methods have been used to show that newborn olfactory neurons form functional synaptic connects with mature neurons (Bardy et al. 2010) and activation of adult-born neurons increases olfactory learning and memory (Alonso et al. 2012).

With an effort to potentially translate the use of optogenetics into clinical therapies, ChR2 has been targeted to photoreceptor-deprived retinas to restore vision in blind mice (Caporale et al. 2011, Doroudchi et al. 2011). In another study, ChR2 was expressed in dorsal root ganglion neurons whose terminals project to the plantar skin. Shining blue LED light on the skin led to touch sensation in these transgenic mice (Ji et al. 2012), suggesting the potential to recover sensory input in clinical conditions of spinal cord damage.

Nonhuman Primate

Expression of opsins in nonhuman primates relies on viral gene delivery methods, so it remains difficult to target specific cells. However, several studies have demonstrated success in driving cortical neurons optogenetically with ChR2, Arch, and Halo (Diester et al. 2011, Han et al. 2009). Most recently, researchers have altered behavioral performance in monkeys with optogenetics. For example, when the neurons in superior colliculus are silenced with optogenetics, there is an increase in the latency of saccadic eye movements and a decrease in the speed of movement (Cavanaugh et al. 2012). On the other hand, stimulating the neurons in arcuate sulcus decreases the delay of the saccadic eye movement (Gerits et al. 2012).

Concluding Remarks

Optogenetics offers new ways to modify neural activity rapidly and reversibly and has been applied to many animal models. Three major classes of light-activated opsins, channelrhodopsins for neural activation and halorhodopsins and archaerhodopsins for neural silencing, can generate sufficient photocurrents in mammalian neurons when irradiated with light. By targeting these opsins to specific cells of interest, it is now possible to investigate the roles of these neurons in neural computation and animal behavior. While optogenetics has revolutionized the investigation of specific cells by providing high spatial, temporal, and cell-type specificity, it is important to properly interpret results from optogenetic perturbation studies. Continued progress in generating novel opsins with improved functions or power spectra, along with increasingly available transgenic animal models and associated hardware, will facilitate the continued widespread use of optogenetics in mapping neural circuits (Table 9.1).

Frequently Asked Questions

Q: What is the difference between halorhodopsin (Halo) and archaerhodopsin (Arch)?

A: Although both Halo and Arch are optical silencers, their mechanisms involve transport of different ions. Halo is a light-activated (inward) chloride pump and Arch is a light-activated (outward) proton pump. Halo activation could potentially alter the gradient between extra- and intracellular chloride concentration, which is critical for the proper function of some ion channels, specifically the GABA receptor. On the other hand, activating Arch for a prolonged period could change the pH of the intra- and extracellular environments.

Q: How does one choose between using viral gene delivery or using transgenic mice?

A: Both methods have their advantages and disadvantages. Opsin expression by viral gene delivery is usually limited to the area around the injection site, so it is preferred if researchers only want to stimulate neurons in a small area. However, virus injections

TABLE 9.1 Necessary Equipment

Equipment	Potential Sources
Light Source: *Both lasers and LEDs can be used as light sources. It is important to make sure that the amplitude of the light output is stable over time.*	
Laser	LaserGlow, Shanghai Laser
LED	Thorlabs
Optical Equipment: *Both lasers and LEDs can be used as light sources. It is important to make sure that the amplitude of the light output is stable over time.*	
Optical fiber	Thorlabs
Ceramic ferrule	Thorlabs
Stainless ferrule	Thorlabs
Ferrule mating sleeves	Thorlabs
Fiber optic rotary joint	Doric Lenses
Transgenic Mice: *An easy way to obtain mice with cell-type–specific opsin expression is to cross the transgenic mice with Cre-driven opsin expression with the mice specifically expressing Cre in the desired cell type.*	
Mice	Jackson Labs
Electrical commutator with optogenetics support: *The TDT commutator provides a single optical fiber with rotary joint, which can be used with any light source, while Plexon incorporates LED into its commutator.*	
Commutator with optical fiber	Tucker-Davis Technologies; Plexon

could damage the brain tissue locally, so using some experimental techniques, such as electrophysiology, to collect data around the injection site may be problematic. In transgenic mice, the opsin expression level is usually higher and therefore less light needs to be used for experiments. The expression is usually specific to the cell type and occurs throughout the whole brain, so it is preferable when one wants to stimulate large brain areas. Unfortunately, this may become a disadvantage if researchers want to study a small brain structure because light illumination of a large volume may stimulate opsin-expressing neurons outside of the target area.

Q: How does one determine the optical fiber size?

A: An optical fiber with a larger diameter can deliver more light and illuminate a larger area. However, a larger fiber also potentially causes more damage to the brain. An alternative way to illuminate a large area is to use multiple small fibers. These small fibers are usually arranged in a specific pattern to cover the desired area.

References

Adamantidis, A., Carter, M. C., & de Lecea L. (2010). Optogenetic deconstruction of sleep- wake circuitry in the brain. *Front Mol Neurosci* 2: 31.

Adamantidis, A. R., Zhang, F., Aravanis, A. M., et al. (2007). Neural substrates of awakening probed with optogenetic control of hypocretin neurons. *Nature* 450: 420–424.

Airan, R. D., Thompson, K. R., Fenno, L. E., et al. (2009). Temporally precise in vivo control of intracellular signalling. *Nature* 458: 1025–1029.

Alonso, M., Lepousez, G., Wagner, S., et al. (2012). Activation of adult-born neurons facilitates learning and memory. *Nat Neurosci* 15: 897–904.

Anikeeva, P., Andalman, A. S., Witten, I., et al. (2012). Optetrode: a multichannel readout for optogenetic control in freely moving mice. *Nat Neurosci* 15: 163–170.

Arrenberg, A. B., Stainier, D. Y., Baier, H., & Huisken, J. (2010). Optogenetic control of cardiac function. *Science* 330: 971–974.

Atasoy, D., Aponte, Y., Su, H. H., & Sternson, S. M. (2008). A FLEX switch targets channelrhodopsin-2 to multiple cell types for imaging and long-range circuit mapping. *J Neurosci* 28: 7025–7030.

Ayling, O. G., Harrison, T. C., Boyd, J. D., et al. (2009). Automated light- based mapping of motor cortex by photoactivation of channelrhodopsin-2 transgenic mice. *Nat Meth* 6: 219–224.

Bamann, C., Gueta, R., Kleinlogel, S., et al. (2010) Structural guidance of the photocycle of channelrhodopsin-2 by an interhelical hydrogen bond. *Biochemistry* 49: 267–278.

Bamberg, E., Tittor, J., & Oesterhelt, D. (1993). Light-driven proton or chloride pumping by halorhodopsin. *Proc Natl Acad Sci USA* 90: 639–643.

Banghart, M., Borges, K., Isacoff, E., et al. (2004). Light-activated ion channels for remote control of neuronal firing. *Nat Neurosci* 7: 1381–1386.

Bardy, C., Alonso, M., Bouthour, W., & Lledo, P. M. (2010). How, when, and where new inhibitory neurons release neurotransmitters in the adult olfactory bulb. *J Neurosci* 30: 17023–17034.

Bellmann, D., Richardt, A., Freyberger, R., et al. (2010). Optogenetically induced olfactory stimulation in Drosophila larvae reveals the neuronal basis of odor-aversion behavior. *Front Behav Neurosci* 4: 27.

Berndt, A., Schoenenberger, P., Mattis, J., et al. (2011). High-efficiency channelrhodopsins for fast neuronal stimulation at low light levels. *Proc Natl Acad Sci USA* 108: 7595–7600.

Berndt, A., Yizhar, O., Gunaydin, L. A., et al. (2009). Bi-stable neural state switches. *Nat Neurosci* 12: 229–234.

Bernstein, J. G., & Boyden, E. S. (2011). Optogenetic tools for analyzing the neural circuits of behavior. *Trends Cogn Sci* 15: 592–600.

Bernstein, J. G., Garrity, P. A., & Boyden, E. S. (2012). Optogenetics and thermogenetics: technologies for controlling the activity of targeted cells within intact neural circuits. *Curr Opin Neurobiol* 22: 61–71.

Bernstein, J. G., Han, X., Henninger, M. A., et al. (2008). Prosthetic systems for therapeutic optical activation and silencing of genetically-targeted neurons. *Proceedings—Society of Photo-Optical Instrumentation Engineers* 6854: 68540H.

Boyden, E. S., Zhang, F., Bamberg, E., et al. (2005). Millisecond-timescale, genetically targeted optical control of neural activity. *Nat Neurosci* 8: 1263–1268.

Bundschuh, S. T., Zhu, P., Scharer, Y. P., & Friedrich, R. W. (2012). Dopaminergic modulation of mitral cells and odor responses in the zebrafish olfactory bulb. *J Neurosci* 32:6830–6840.

Busskamp, V., Picaud, S., Sahel, J. A., & Roska, B. (2012). Optogenetic therapy for retinitis pigmentosa. *Gene Ther* 19: 169–175.

Caporale, N., Kolstad, K. D., Lee, T., et al. (2011). LiGluR restores visual responses in rodent models of inherited blindness. *Mol Ther* 19: 1212–1219.

Cardin, J. A., Carlen, M., Meletis, K., et al. (2009). Driving fast-spiking cells induces gamma rhythm and controls sensory responses. *Nature* 459: 663–667.

Cavanaugh, J., Monosov, I. E., McAlonan, K., et al. (2012). Optogenetic inactivation modifies monkey visuomotor behavior. *Neuron* 76: 901–907.

Chow, B. Y., Han, X., & Boyden, E. S. (2012). Genetically encoded molecular tools for light- driven silencing of targeted neurons. *Progr Brain Res* 196: 49–61.

Chow, B. Y., Han, X., Dobry, A. S., et al. (2010). High-performance genetically targetable optical neural silencing by light-driven proton pumps. *Nature* 463: 98–102.

Cruikshank, S. J., Urabe, H., Nurmikko, A. V., & Connors, B. W. (2010). Pathway-specific feedforward circuits between thalamus and neocortex revealed by selective optical stimulation of axons. *Neuron* 65: 230–245.

Diester, I., Kaufman, M. T., Mogri, M., et al. (2011). An optogenetic toolbox designed for primates. *Nat Neurosci* 14: 387–397.

Doroudchi, M. M., Greenberg, K. P., Zorzos, A. N., et al. (2011). Towards optogenetic sensory replacement. *Conference Proceedings: Annual International Conference of the IEEE Engineering in Medicine and Biology Society IEEE Engineering in Medicine and Biology Society Conference* 2011: 3139–3141.

Drew, P. J., Shih, A. Y., Driscoll, J. D., et al. (2010). Chronic optical access through a polished and reinforced thinned skull. *Nat Meth* 7: 981–984.

Duschl, A., Lanyi, J. K., & Zimanyi, L. (1990). Properties and photochemistry of a halorhodopsin from the haloalkalophile, *Natronobacterium pharaonis. J Biol Chem* 265: 1261–1267.

Feldbauer, K., Zimmermann, D., Pintschovius, V., et al. (2009). Channelrhodopsin-2 is a leaky proton pump. *Proc Natl Acad Sci USA* 106: 12317–12322.

Gerits, A., Farivar, R., Rosen, B. R., et al. (2012). Optogenetically induced behavioral and functional network changes in primates. *Curr Biol* 22: 1722–1726.

Govorunova, E. G., Spudich, E. N., Lane, C. E., et al. (2011). New channelrhodopsin with a red-shifted spectrum and rapid kinetics from *Mesostigma viride. MBio* 2: e00115–00111.

Gradinaru, V., Thompson, K. R., & Deisseroth, K. (2008). eNpHR: a Natronomonas halorhodopsin enhanced for optogenetic applications. *Brain Cell Biol* 36: 129–139.

Gradinaru, V., Zhang, F., Ramakrishnan, C., et al. (2010). Molecular and cellular approaches for diversifying and extending optogenetics. *Cell* 141: 154–165.

Gunaydin, L. A., Yizhar, O., Berndt, A., et al. (2010). Ultrafast optogenetic control. *Nat Neurosci* 13: 387–392.

Han, X. (2012a). In vivo application of optogenetics for neural circuit analysis. *ACS Chem Neurosci* 3: 577–584.

Han, X. (2012b). Optogenetics in the nonhuman primate. *Progr Brain Res* 196: 215–233.

Han, X., & Boyden, E. S. (2007). Multiple-color optical activation, silencing, and desynchronization of neural activity, with single-spike temporal resolution. *PLoS One* 2: e299.

Han, X., Chow, B. Y., Zhou, H., et al. (2011). A high-light sensitivity optical neural silencer: development and application to optogenetic control of non-human primate cortex. *Front Syst Neurosci* 5: 18.

Han, X., Qian, X., Bernstein, J. G., et al. (2009). Millisecond-timescale optical control of neural dynamics in the nonhuman primate brain. *Neuron* 62: 191–198.

Honjo, K., Hwang, R. Y., & Tracey, W. D. Jr. (2012). Optogenetic manipulation of neural circuits and behavior in Drosophila larvae. *Nat Protocols* 7: 1470–1478.

Husson, S. J., Costa, W. S., Wabnig, S., et al. (2012). Optogenetic analysis of a nociceptor neuron and network reveals ion channels acting downstream of primary sensors. *Curr Biol* 22: 743–752.

Inada, K., Kohsaka, H., Takasu, E., et al. (2011). Optical dissection of neural circuits responsible for Drosophila larval locomotion with halorhodopsin. *PLoS One* 6: e29019.

Janovjak, H., Szobota, S., Wyart, C., et al. (2010). A light-gated, potassium-selective glutamate receptor for the optical inhibition of neuronal firing. *Nat Neurosci* 13: 1027–1032.

Ji, Z. G., Ito, S., Honjoh, T., et al. (2012). Light-evoked somatosensory perception of transgenic rats that express channelrhodopsin- 2 in dorsal root ganglion cells. *PLoS One* 7: e32699.

Kawazoe, Y., Yawo, H., & Kimura, K. D. (2012). A simple optogenetic system for behavioral analysis of freely moving small animals. *Neurosci Res* 75: 65–68.

Kleinlogel, S., Feldbauer, K., Dempski, R. E., et al. (2011). Ultra light-sensitive and fast neuronal activation with the Ca(2)+ permeable channelrhodopsin CatCh. *Nat Neurosci* 14: 513–518.

Knopfel, T., & Boyden, E. S. (Eds.) (2012). *Optogenetics: Tools for controlling and monitoring neuronal activity.* Elsevier, Amsterdam.

Kravitz, A. V., & Kreitzer, A. C. (2011). Optogenetic manipulation of neural circuitry in vivo. *Curr Opin Neurobiol* 21: 433–439.

Letzkus, J. J., Wolff, S. B., Meyer, E. M., et al. (2011). A disinhibitory microcircuit for associative fear learning in the auditory cortex. *Nature* 480: 331–335.

Li, X., Gutierrez, D. V., Hanson, M. G., et al. (2005). Fast noninvasive activation and inhibition of neural and network activity by vertebrate rhodopsin and green algae channelrhodopsin. *Proc Natl Acad Sci USA* 102: 17816–17821.

Lima, S. Q., & Miesenbock, G. (2005). Remote control of behavior through genetically targeted photo-stimulation of neurons. *Cell* 121: 141–152.

Lin, J. Y., Lin, M. Z., Steinbach, P., & Tsien, R. Y. (2009). Characterization of engineered channelrhodopsin variants with improved properties and kinetics. *Biophys J* 96: 1803–1814.

Luo, D., & Saltzman, W. M. (2000). Synthetic DNA delivery systems. *Nature Biotechnology* 18: 33–37.

Ma, D., Zerangue, N., Lin, Y. F., et al. (2001). Role of ER export signals in controlling surface potassium channel numbers. *Science* 291: 316–319.

Mazarakis, N. D., Azzouz, M., Rohll, J. B., et al. (2001). Rabies virus glycoprotein pseudotyping of lentiviral vectors enables retrograde axonal transport and access to the nervous system after peripheral delivery. *Hum Mol Genet* 10: 2109–2121.

Miesenbock, G. (2011) Optogenetic control of cells and circuits. *Annu Rev Cell Dev Biol* 27: 731–758.

Mobley, J., & Vo-Dinh, T. (2003). Optical properties of tissue. In T. Vo-Dinh (Ed.), *Biomedical photonics handbook* (pp. 1–72). Boca Raton, FL: CRC Press.

Nagel, G., Brauner, M., Liewald, J. F., et al. (2005). Light activation of channelrhodopsin-2 in excitable cells of *Caenorhabditis elegans* triggers rapid behavioral responses. *Curr Biol* 15: 2279–2284.

Nagel, G., Szellas, T., Huhn, W., et al. (2003). Channelrhodopsin-2, a directly light-gated cation-selective membrane channel. *Proc Natl Acad Sci USA* 100: 13940–13945.

Okazaki, A., Sudo, Y., & Takagi, S. (2012). Optical silencing of *C. elegans* cells with arch proton pump. *PLoS One* 7: e35370.

Petreanu, L., Huber, D., Sobczyk, A., & Svoboda, K. (2007). Channelrhodopsin-2-assisted circuit mapping of long-range callosal projections. *Nat Neurosci* 10: 663–668.

Raimondo, J. V., Kay, L., Ellender, T. J., & Akerman, C. J. (2012). Optogenetic silencing strategies differ in their effects on inhibitory synaptic transmission. *Nat Neurosci* 15:1102–1104.

Rolls, A., Colas, D., Adamantidis, A., et al. (2011). Optogenetic disruption of sleep continuity impairs memory consolidation. *Proc Natl Acad Sci USA* 108: 13305–13310.

Sauer, B., & Henderson N. (1988). Site-specific DNA recombination in mammalian cells by the Cre recombinase of bacteriophage P1. *Proc Natl Acad Sci USA* 85: 5166–5170.

Sineshchekov, O. A., Govorunova, E. G., & Spudich, J. L. (2009). Photosensory functions of channelrhodopsins in native algal cells. *Photochem Photobiol* 85: 556–563.

Sohal, V. S., Zhang, F., Yizhar, O., & Deisseroth, K. (2009). Parvalbumin neurons and gamma rhythms enhance cortical circuit performance. *Nature* 459: 698–702.

Spudich, J. L., Yang, C. S., Jung, K. H., & Spudich, E. N. (2000). Retinylidene proteins: structures and functions from archaea to humans. *Annu Rev Cell Dev Biol* 16: 365–392.

Stark, E., Koos, T., & Buzsaki, G. (2012). Diode probes for spatiotemporal optical control of multiple neurons in freely moving animals. *J Neurophysiol* 108: 349–363.

Stirman, J. N., Crane, M. M., Husson, S. J., et al. (2011). Real-time multimodal optical control of neurons and muscles in freely behaving *Caenorhabditis elegans*. *Nat Meth* 8: 153–158.

Timbers, T. A., Giles, A. C., Ardiel, E. L., et al. (2012). Intensity discrimination deficits cause habituation changes in middle-aged *Caenorhabditis elegans*. *Neurobiol Aging* 34: 621–631.

Tsien, J. Z., Chen, D. F., Gerber, D., et al. (1996). Subregion- and cell type-restricted gene knockout in mouse brain. *Cell* 87: 1317–1326.

Tufail, Y., Yoshihiro, A., Pati, S., et al. (2011). Ultrasonic neuromodulation by brain stimulation with transcranial ultrasound. *Nat Protocols* 6: 1453–1470.

Tye, K. M., Prakash, R., Kim, S.Y., et al. (2011). Amygdala circuitry mediating reversible and bidirectional control of anxiety. *Nature* 471: 358–362.

Voigts, J., Siegle, J. H., Pritchett, D. L., & Moore, C. I. (2013). The flexDrive: an ultra-light implant for optical control and highly parallel chronic recording of neuronal ensembles in freely moving mice. *Front Syst Neurosci* 7:8.

Waehler, R., Russell, S. J., & Curiel, D. T. (2007). Engineering targeted viral vectors for gene therapy. *Nat Rev Genet* 8: 573–587.

Wang, H., Sugiyama, Y., Hikima, T., et al. (2009). Molecular determinants differentiating photocurrent properties of two channelrhodopsins from chlamydomonas. *J Biol Chem* 284: 5685–5696.

Wang, J., Ozden, I., Diagne, M., et al. (2011). Approaches to optical neuromodulation from rodents to non-human primates by integrated optoelectronic devices. *Conference Proceedings: Annual International Conference of the IEEE Engineering in Medicine and Biology Society IEEE Engineering in Medicine and Biology Society Conference* 2011: 7525–7528.

Wang, J., Wagner, F., Borton, D. A., et al. (2012). Integrated device for combined optical neuromodulation and electrical recording for chronic in vivo applications. *J Neural Eng* 9: 016001.

Wells, J., Kao, C., Mariappan, K., et al. (2005). Optical stimulation of neural tissue in vivo. *Opt Lett* 30: 504–506.

Wen, L., Wang, H., Tanimoto, S., et al. (2010). Opto-current-clamp actuation of cortical neurons using a strategically designed channelrhodopsin. *PLoS One* 5: e12893.

Wentz, C. T., Bernstein, J. G., Monahan, P., et al. (2011). A wirelessly powered and controlled device for optical neural control of freely- behaving animals. *J Neural Eng* 8: 046021.

Wickersham, I. R., Sullivan, H. A., & Seung, H. S. (2010). Production of glycoprotein-deleted rabies viruses for monosynaptic tracing and high-level gene expression in neurons. *Nat Protocols* 5: 595–606.

Witten, I. B., Lin, S. C., Brodsky, M., et al. (2010). Cholinergic interneurons control local circuit activity and cocaine conditioning. *Science* 330: 1677–1681.

Ye, H., Daoud-El Baba, M., Peng, R. W., & Fussenegger, M. (2011). A synthetic optogenetic transcription device enhances blood-glucose homeostasis in mice. *Science* 332: 1565–1568.

Yizhar, O., Fenno, L. E., Davidson, T. J., et al. (2011a). Optogenetics in neural systems. *Neuron* 71: 9–34.

Yizhar, O., Fenno, L. E., Prigge, M., et al. (2011b). Neocortical excitation/inhibition balance in information processing and social dysfunction. *Nature* 477: 171–178.

Zemelman, B. V., Lee, G. A., Ng, M., & Miesenbock, G. (2002). Selective photostimulation of genetically charged neurons. *Neuron* 33: 15–22.

Zemelman, B. V., Nesnas, N., Lee, G. A., & Miesenbock, G. (2003). Photochemical gating of heterologous ion channels: remote control over genetically designated populations of neurons. *Proc Natl Acad Sci USA* 100: 1352–1357.

Zhang, F., Gradinaru, V., Adamantidis, A. R., et al. (2010). Optogenetic interrogation of neural circuits: technology for probing mammalian brain structures. *Nat Protocols* 5: 439–456.

Zhang, F., Prigge, M., Beyriere, F., et al. (2008). Red-shifted optogenetic excitation: a tool for fast neural control derived from *Volvox carteri*. *Nat Neurosci* 11: 631–633.

Zhang, F., Vierock, J., Yizhar, O., et al. (2011). The microbial opsin family of optogenetic tools. *Cell* 147: 1446–1457.

Zhang, F., Wang, L. P., Brauner, M., et al. (2007). Multimodal fast optical interrogation of neural circuitry. *Nature* 446: 633–639.

Zhao, S., Cunha, C., Zhang, F., et al. (2008). Improved expression of halorhodopsin for light-induced silencing of neuronal activity. *Brain Cell Biol* 36: 141–154.

Zhu, P., Fajardo, O., Shum, J., et al. (2012). High-resolution optical control of spatiotemporal neuronal activity patterns in zebrafish using a digital micromirror device. *Nat Protocols* 7: 1410–1425.

Zorzos, A. N., Boyden, E. S., & Fonstad, C. G. (2010). Multiwaveguide implantable probe for light delivery to sets of distributed brain targets. *Opt Lett* 35: 4133–4135.

Zorzos, A. N., Dietrich, A., Talei Franzesi, G., et al. (2009). Light-proof neural recording electrodes. Society for Neuroscience meeting. Program No. 388.12/GG107

Glossary

Adsorption. The adhesion of molecules in a liquid solution, gas, or dissolved solid to a solid surface.

ACSF. Artificial Cerebrospinal fluid.

Amplifier. An electronic unit that connects to a headstage, and provides filtering, current and voltage command adjustments, and electrode resistance compensation. The amplifier may be computer-controlled or may only provide analog signals to the computer.

Anodic voltage sweep. The phase of the voltammetric waveform during which the voltage is linearly increased from the holding potential to the maximum potential. During this time electrons flow from the electrode surface into chemical species surrounding the electrode.

Archaerhodopsins. A group of light-activated proton pumps. A specific type of Archaerhodopsin, Archaerhodopsin-3 (Arch) from *Halorubrum sodomense*, can be used as an optical silencer with peak activation wavelength at 566 nm.

Artifacts. Non-experimental events contributing to recordings of electrical activity, which can be biological (e.g. intrinsic to the preparation or subject) or environmental (e.g., power line noise).

Bandpass filter. Attenuates electrical activity outside of predetermined minimum and maximum frequency band values.

Capacitance. The ability to store an electrical charge across an insulator. Units are Farads (F), or divisions thereof (i.e., milliFarads). Capacitance effectively filters rapid voltage changes. Neuronal membranes have capacitance, as do glass recording pipettes.

Cathodic voltage sweep. Phase of the voltammetric waveform during which the voltage is linearly decreased from the maximum potential back down to the holding potential. During this time electrons flow from the chemical species surrounding the electrode into the electrode.

Channelrhodopsins. A group of light-gated ion channels. Channelrhodopsin-2 (ChR2) from *Chlamydomonas reinhardtii* is commonly used as an optical activator with peak activation wavelength at 470 nm.

Chemometrics. A set of mathematical tools used to extract relevant information from any potential noise within large data sets. One approach utilizes principal component analysis and regression to statistically identify phasic dopamine events within large data sets.

Chronic microsensor. A chronically implantable working electrode from which recordings can be made throughout experiments that last from weeks to months.

Conductance. The opposite of resistance, the permissivity of a material to current flow. Units are Siemens (S), or divisions thereof (i.e., milliSiemens).

Current. The flow of positive charge. Physiologically, charge is carried by the flow of ions across neuronal membrane. Units are amperes (A), or divisions thereof (i.e., milliamps).

Dissection buffer. A variation of ACSF that is used during tissue dissection and cutting. May have substitutions of ions (e.g., NMDG for sodium chloride), and different pH buffer systems.

Electrical grounding. A direct physical connection between an electrical circuit and earth or some other infinite sink for charge which can absorb an unlimited amount of current without changing its potential. This connectivity serves to minimize the buildup of static electricity, which can inject noise into electrical systems.

Electrolyte. A substance that will ionize in solution (i.e. KCl in water dissociates into K+ and Cl-).

Extracellular recording. A technique that allows for detection of electrical events from the extracellular space proximal to neurons. The tip of the electrode must be close to the neuron of interest to achieve signal isolation, but does not require piercing or disrupting the membrane.

Faradaic current. The current generated by electron flow as a result of oxidation or reduction of a chemical species at an electrode.

Ground loop. An occurrence in an electrical circuit where parts of the circuit have different ground potentials—this can introduce noise from one part of the circuit into another.

Halorhodopsins. A group of light-activated chloride pumps. Halo3.0/eNpHR3.0, which is a recombination of a membrane-targeting sequence and the original Halo, can be used as an optical silencer with peak activation wavelength at 590nm.

Headcap. The cement "hat" that is created on an animal's skull to encase the implanted working and reference electrodes and the wires that link them to a connector that allows for them to be plugged into a headstage within a behavioral chamber.

Headstage. The portion of the amplifier that is placed close to the preparation and usually also holds the electrode.

Electrode solution. See internal solution

Electroencephalogram (EEG). The recording of electrical activity on the scalp.

EEG coherence. The magnitude of the coupling between two different time series (e.g. from two electrodes) independent of their power.

EEG power. Power within specified frequency bands indexes the average magnitude of oscillations over a specified time range.

ERP component. A positive- or negative-going deflection in the ERP waveform occurring with specific latencies relative to the stimulus. Examples include the N100 or P300—the letter indicates the polarity of the deflection (positive or negative) and the number indicates the approximate timing of the deflection (onset or peak, post stimulus) in milliseconds.

Event code. Trigger signals that reflect events of interest are sent to the recording software as they occur so that the neural response can be time-locked to the event during signal averaging.

Event-related potentials (ERPs). Averaged electrical activity time-locked to a specific event that occurs repeatedly. ERPs can be characterized by their shape, magnitude, and latency.

Independent Component Analysis (ICA). A blind source separation method that transforms signal mixtures into components from their original sources. It can be used for spike-sorting—extracting single neuron action potential traces from the complex raw PDA recordings, where many diodes record multiple neurons, and many neurons are redundantly recorded by many diodes.

Inion. The lowest point of the skull at the back of the head, a reference point typically used for standard placement of the EEG cap.

Internal solution. The solution that is used in a patch pipette, usually similar to the internal salt contents of a cell.

Invertebrate ganglia. Invertebrate nervous systems are organized into ganglia, each consisting of dozens to a few thousand individual neurons. The relatively small number and large size of the neurons in each ganglion makes these preparations attractive as model systems for studying the neural basis of behavior.

Liquid junction potential. A voltage that forms when two conductors of different composition come into contact. For example, junction potentials occur when two solutions that have different ionic concentrations come in contact with each other. In the context of extracellular recording, this can occur when the electrolyte fluid inside the recording pipette is not the same as the recording bath (or extracellular fluid), or between the electrode solution and its wire. Junction potential can cause measurement errors and should be offset mathematically or electronically.

Low-pass filter. A filter that allows for the removal of high-frequency noise from electrical signals.

Montage. The arrangement of EEG channels, with the most common choice being the referential montage whereby each recorded channel reflects the difference between that electrode site and a reference electrode site.

Nasion. The point between the forehead and the nose, a reference point typically used for standard placement of the EEG cap.

Opsins. A group of light sensitive proteins some of which are ion channels while other are ion pumps. Light-gated ion channels use the energy from light to open the ion channel, allowing select ions pass through the channel along their concentration gradient. Light-activated pumps use the energy from light to actively move the ions across the cell membrane.

Optogenetics. A technique allows fast and reversible activation or silencing of neurons with light. In general, it involves two steps: genetically modifying cells with an opsin to render them light-sensitive, and using light to alter the activity of these cells.

Oxidation/Reduction potential. The voltage at which a molecule undergoes oxidation or reduction reaction.

Patch Pipette. A small diameter glass capillary that has been pulled to have a rapid taper to a small tip (1–2 μm diameter) that is fire polished.

Photodiode Array (PDA). A densely packed array of light sensors, which can be used with a voltage-sensitive dye to detect neural activity at hundreds of locations simultaneously.

Pyramidal cells. A type of neuron characterized by multiple dendrites and a single axon, whose integrated activity contributes to the activity recorded by EEG electrodes.

Recording chamber. A polycarbonate or plexiglass chamber on the microscope stage, usually with a glass bottom. The chamber can be heated, has inlet and outlet connections for solution exchange, and provisions for positioning a reference electrode.

Redox (oxidation/reduction) reaction. A chemical reaction in which a molecule's oxidation state is changed, that is, the molecule undergoes a loss (oxidation) or gain (reduction) of electrons. The ease of occurrence of these reactions depends upon the molecular structure, thus different molecules will undergo oxidation and reduction at different voltages.

Reference electrode. An electrode having a stable and well-known electrical potential against which the potential at the working electrode can be referenced.

Resistance. A measure of opposition to current flow. Also called "impedance." Units are ohms(Ω), or divisions thereof (i.e., milliohms).

Single unit. A "unit" is an empirically based assessment of a physiological waveform characterized by consistent height and shape and does not sum with itself (that is it has a refractory period). The experimenter's implicit assumption is that a unit represents the activity of an individual neuron.

Spiral waves. A rotating wave pattern, which is commonly seen in dynamic systems such as storms, flowing rivers, fibrillating heart and oscillating brain cortex. A spiral originates from a "central rotor". Continuous rotating of the rotor generates an outward wave that swipes across the space. The center of the spiral rotor is often referred to as a "phase singularity", in which all phases (between 0–2pi) are present in a single point.

Voltage. A difference in electrical charge between two points, such as across a neuronal membrane. Also called a "potential." Units are volts (V), or divisions thereof (i.e., millivolts).

Voltage-sensitive dye imaging. When neurons are stained with a voltage-sensitive dye, they flicker in brightness as they fire action potentials, making it possible to record their activity with a PDA.

Working electrode. An electrode in an electrochemical system at which the reaction of interest occurs (i.e. detection of dopamine).

Index